Future Predictions By an Engineer and Seer

This book is about predictions of the Future from both projections of existing trends and possible major paradigm changes.
On Making Predictions of the Future:

It is possible to make pretty good future predictions by taking into account three things--

1) Technological and Sociological Trends
2) Potential Paradigm changes
3) Intuition

In this book you will see many technological trends analyzed by an Engineer and using Intuition to help pick the most likely trends in our future.

What are possible major paradigm changes which will also affect our future? Projections of the year 2100 as well as up to 1,000 years ahead.

My goal in this book is to make the most accurate predictions of how our future will be over the next centuries that has ever been tried.

What does the Future Hold?

What will the world look like in the future?

How will we live our daily lives?

Future Predictions By an Engineer and Seer

Copyright Page

Future Predictions by an Engineer and Seer

Copyrighted By Martin K. Ettington 2018 USA

All Rights Reserved

ISBN: 9781980312284

Future Predictions By an Engineer and Seer

Future Predictions By an Engineer and Seer

Other books by Martin K. Ettington

Spiritual and Metaphysics Books:
Prophecy: A History and How to Guide
God Like Powers and Abilities
Enlightenment for Newbies
Removing Illusions to Find True Happiness
Using the Scientific Method to Study the Paranormal
A Compendium of Metaphysics and How to Guides (Six books together in one volume)
Love from the Heart
The Enlightenment Experience
Learn Your Soul's Purpose
Pursuing Enlightenment
A Modern Man's Search for Truth
Use Intuition and Prophecy to Improve Your Life
The Handbook of Spiritual and Energy Healing

Longevity & Immortality:
Physical Immortality: A History and How to Guide
The Commentaries of Living Immortals
Records of Extremely Long Lived Persons
Enlightenment and Immortality
Longevity Improvements from Science
The 10 Principles of Personal Longevity
Telomeres & Longevity
The Diets and Lifestyles of the Worlds Oldest Peoples
The Longevity Six Books Bundle

Science Fiction:
Out of This Universe
Personal Freedom-Parts 1 & 2
The Psychic Soldier Series:
   Book 1-Himalayan Journey
   Book 2-A Soldier is Born
   Book 3-Fighting For Right
   Book 4-Earth Protector
The Immortality Sci Fi Bundle

The God Like Powers Series:
Human Invisibility
Invulnerability and Shielding
Teleportation
Psychokinesis
Our Energy Body, Auras, and Thoughtforms

The God Like Powers Series—
   Volume 1 Compilation
The Yoga Discovery Series:
Yoga-An Ancient Art Form
Hatha Yoga-Helping you Live Better
Raja Yoga-Through the Ages
The Yoga Discovery Package

Business & Coaching Books:
Creating, Publishing, & Marketing Practitioner Ebooks
Building a Successful Longevity Coaching Business
Why Become a Coach?
The Professional Coaching Success Trilogy
2020-Make Money Writing and Selling Books
The 2020 Handbook of High Paying Work Without a College Degree

Science, Technology, and Misc.
Future Predictions By and Engineer & Seer
The Unusual Science & Technology Bundle
The Real Atlantis-In the Eye of the Sahara
Are Cryptozoological Animals Real or Imaginary?
Real Time Travel Stories From a Psychic Engineer
Removing Limits On Our Consciousness- And Thinking Outside the Box
33 Incredible True Survival Stories
How to Survive Anything: From the Wilderness to Man Made Disasters
All About Mars Journeys and Settlement
Mining the Asteroid Belt

Ancient History
The Real Atlantis-In the Eye of the Sahara
Ancient & Prehistoric Civilizations
Ancient & Prehistoric Civilizations-Book Two
The History of Antediluvian Giants
The Antediluvian History of Earth
Ancient Underground Cities and Tunnels
Strange Objects Which Should Not Exist
Strange and Ancient Places in the USA
A Theory of Ancient Prehistory And Giant Aliens

Aliens and Space
Aliens and Secret Technology
Aliens Are Already Among Us

Future Predictions By an Engineer and Seer

Designing and Building Space Colonies
Humanity and the Universe
All About Moon Bases
All About Mars Journeys and Settlement
The Space and Aliens Six Books Bundle

A Theory of Ancient Prehistory and Giant Aliens
The Space Colonies and Space Structures Coloring Book
All About Asteroids

The Longevity Training Series

(A transcription of the online Multimedia Longevity Coaching Training Program)

The Personal Longevity Training Series-Book1-Long Lived Persons
The Personal Longevity Training Series-Book2-Your Soul's Purpose
The Personal Longevity Training Series-Book3-Enable Your Life Urge
The Personal Longevity Training Series-Book4-Your Spiritual Connection
The Personal Longevity Training Series-Book5-Having Love in Your Heart
The Personal Longevity Training Series-Book6-Energy Body Health
The Personal Longevity Training Series-Book7-The Science of Longevity
The Personal Longevity Training Series-Book8-Physical Body Health
The Personal Longevity Training Series-Book9-Avoiding Accidents
The Personal Longevity Training Series-Book10-Implementing These Principles

The Personal Longevity Training Series-Books One Thru Ten

These books are all available in digital and printed formats from my website and on Amazon, Barnes & Noble, Apple ITunes, and many other sites

My Books Website is: http://mkettingtonbooks.com

Signup for our Mailing List to get the following:

1) A discount coupon for 25% discount on all books on our site

2) Occasional Notices of new books available

3) Occasional Email on other offerings of ours (Monthly)

Go to this link to sign-up:

http://personal-longevity.com/mkebooks/emailsignup/

And click this link to get the FREE 102 page Ebook titled "Secrets of Many Things"

If you have any questions about this book or other subjects please contact the Author at:

mke@mkettingtonbooks.com

Future Predictions By an Engineer and Seer

Table of Contents

1.0 Introduction 11
Part 1—The History & Difficulty of Predictions 13
2.0 The Usefulness of Good Predictions 13
3.0 Wrong or Limited Insights 15
   3.1 Global Warming Mistakes 17
   3.2 Space Travel after Moon Landings 19
   3.3 Not seeing the possibilities 21
4.0 Previous Prediction Theories and Books 25
   4.1 Malthusian Theory 25
   4.2 Future Shock 27
   4.3 The Third Wave 29
   4.4 Megatrends 31
5.0 Science Fiction Predictions 33
   5.1 Jules Verne 33
   5.2 Arthur C. Clarke 35
   5.3 H.G. Wells 38
   5.4 Robert Heinlein 43
   5.5 Misc Science Fiction Writers 47
6.0 The Singularity 49
7.0 Secret Science and Knowledge 53
8.0 Types of Predictions 55
   8.1 Predictions based on existing trends 55
   8.2 Potential Paradigm shifts 57
   8.3 Predictions based on Intuition & Prophecy 61
   8.4 Taking into Account Economics and Industry 65

9.0 Mixing Logic and Knowledge with Intuition 67

Part 2-Major Trends & Predictions 69

10.0 World Population Growth 69

11.0 Availability of Resources 71

12.0 Technology Development Trends 73

    12.1 Artificial Intelligence 73

    12.2 Biology & DNA 75

    12.3 Construction 77

    12.4 Crypto Currencies 79

    12.5 Drone Technology 81

    12.6 Education 83

    12.7 Farming 85

    12.8 Electrical Power Generation 87

    12.9 Human Longevity 91

    12.10 Manufacturing 95

    12.11 Medicine 97

    12.12 Mobile Technologies 99

    12.13 Military Technology 101

    12.14 Nanotechnology 105

    12.15 Quantum Technologies 107

    12.16 Robotics 109

    12.17 Space Travel & Space Settlements 111

    12.18 Transportation 115

13.0 Human Intelligence 119

14.0 Trends in Human Interactions 123

    14.1 Overall Trends in Work and Jobs 123

    14.2 Trends in Corporations 125

14.3 Trends in Government 127

14.4 Trends in Religious and Spiritual Areas 129

14.5 Trends in Social Interaction and Relationships 131

15.0 Paradigm Shifting 133

15.1 Summary of Paradigm Shift Potentials 135

16.0 The Speed of Change 139

17.0 Predictions About Human Civilization 141

17.1 In the Year 2100 AD—Roughly 80 Years 143

17.2 In the Year 2250 AD—Roughly 232 Years 147

17.3 In the Year 3000 AD—Roughly 980 Years 149

18.0 Summary 151

Bibliography 153

Future Predictions By an Engineer and Seer

1.0 Introduction

I want to welcome you to this book about future predictions. I'm an engineer, psychic, and have written many books over the years about the spiritual, paranormal, longevity, science fiction, and some science books.

As you can tell I have very wide ranging interests. Having read a couple thousand science fiction and other books in my life, I have a real passion about what our future holds.

Over the years I've read many of the popular futurist books like "Future Shock" and "Megatrends" and I wondered if I could do any better.

What I thought I would do in this book is to examine the history of conventional predictions about the future and try to use my own unique approach to see how well I can do.

What I do very well is to integrate knowledge from other people and produce unique insights into that information.

I also have a lot of experiences with Prophecy as I wrote in my book "Prophecy: A History and How to Guide" where I reviewed the history of prophecy and my own experiences.

In this book it will be interesting to combine the logic on scientific forecasting with my intuitional insights.

So the goal of this book will be to review many previous forecasting models, trend information, and use those plus my intuition to project some future probable scenarios.

I believe it is possible to make the most accurate predictions of the future of humanity ever accomplished by taking into account three things--

1) Technological and Sociological Trends
2) Potential Paradigm changes
3) Intuition

These three dimensions of my approach to future predictions should produce a very innovative look at the future.

Hope you enjoy the ride!

Part 1—The History & Difficulty of Predictions

2.0 The Usefulness of Good Predictions

Man is obsessed about predictions of the future. Since ancient times man has turned to Seers and Prophets to advise leaders and individuals on what the future holds.

There is of course something to be said about the advantages of knowing the future—it helps you plan what you will do, and reduces uncertainty and stress about the future.

In recent times seriously recognized predictions have been made based on logic and an understanding of trends and research in our modern society.

This helps explain the wide popularity of books like "Future Shock" and "Megatrends" in recent decades.

3.0 Wrong or Limited Insights

Many predictions have of course been made which never came true or came true in totally different ways.

One funny one from one hundred years ago (1) was the idea that a bus would transport us under the ocean with a whale harnessed like a horse moving it.

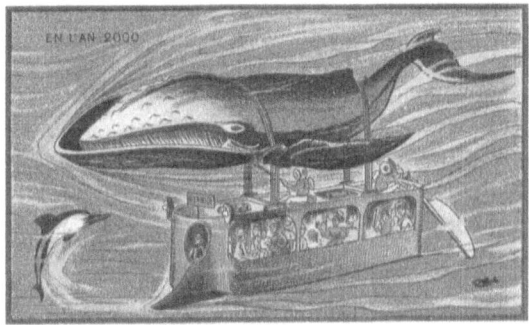

A Whale-Bus

Another prediction from the same article said that there would be aerial firemen putting out fires from above.

Aerial Firemen

3.1 Global Warming Mistakes

There is no doubt that the Earth is still warming from the last Ice Age. Sea levels are estimated to be 120 meters higher since the end of the last Ice Age. Also, the Earth continues to warm and it's estimated the sea level has risen four to eight inches in the last one hundred years.

I remember back in the 1980s when all the scientists were worried about manmade global cooling. There were even some movies from that era showing the Earth freezing to death.

The big issue is how much does man contribute to Global Warming. The scientific evidence is that this issue has been way overblown. As an engineer I've read a lot of the science and found concerns about manmade global warming to be more religious than science based.

Some examples of the flaws in global warming:

1) Predictive computer models which have shown high levels of Earth warming over the last twenty years—which never happened.
2) NOAA the United States National Oceanic and Air Administration has been found to be changing baseline temperature data over decades to show warming trends which were originally not there.
3) Polar Bears were supposed to be dying from melting Arctic Ice several years ago—until the ice was shown to not be melting and the Polar Bears are doing fine.
4) The polar caps are not melting. In fact recent evidence is that they are maintaining their coverage. Some years more and some years less.

Many scientists now believe that long term Earth climate is based on sunspots cycles and they worry we are heading into a solar minimum and global cooling period.

People are often like lemmings and follow the thought leaders like that nutty guy Al Gore who was always looking for a movement to lead but who also becomes less and less logical every year. He keeps proclaiming global warming with no evidence.

It just goes to show that predictions about the future are often based on the current politics of the day, not on facts and scientific analysis.

My prediction based on a lot of the latest scientific evidence is that manmade Global Warming does not exist. That carbon dioxide is such a small component of the atmosphere that it can't affect global temperatures.

A much more likely driver are long term cycles in Sunspots. (2)

My opinion is that manmade Global Warming (also called Anthropomorphic Global Warming) is more of an example of a hysterical prediction of the future than what science really knows.

3.2 Space Travel after Moon Landings

I grew up in the 1960s and remember the moon landings. I used to watch them all on television.

At the time everyone thought we would be living in space and going to Mars by the end of the century in 1999.

What wasn't clear to everyone was how much of our space program then was only based on the cold war and a race for prestige between the United States and the Soviet Union.

I even remember a scene during the movie 2001: A Space Odyssey in 1968 where a Pan Am space shuttle was taking passengers to a large rotating space station in orbit around the Earth. Of course Pan Am which was the largest international carrier then and is now out of business.

Today, the reality is that we just have an orbiting smaller space station and private spaceships are just on the verge of going commercial.

So this again goes to show how strong current beliefs in the future might be totally unfounded since many of these beliefs are based on mistaken understandings of current trends and not a realistic view of what the future holds.

Future Predictions By an Engineer and Seer

3.3 Not seeing the possibilities

Many people who are corporate leaders and/or know powerful technologies intimately still lack the vision as to how we might progress. Here are seven really awful predictions I found from the recent past:

(3) _Foolish Tech Prediction 1_

"I think there is a world market for maybe five computers."

Thomas Watson, president of IBM, 1943

At the dawn of the computer industry, nobody really knew where this new technology would take us. But the explosion of desktop computing that put a PC in nearly every American home within 50 years seems to have eluded the imagination of most mid-century futurists.

After all, when IBM's Thomas Watson said "computer," he meant "vacuum-tube-powered adding machine that's as big as a house." It's fair to say that few people ever wanted one of those, regardless of the size of their desk.

Foolish Tech Prediction 2

"Television won't be able to hold on to any market it captures after the first six months. People will soon get tired of staring at a plywood box every night."

Darryl Zanuck, executive at 20th Century Fox, 1946

By 1946, movie executive Darryl Zanuck had already cemented his place in entertainment history as the producer of more than 100 films for the big silver screen. So who could have blamed him for underestimating the power of the small blue screen? I'm guessing that if Zanuck were alive today, he'd find himself just as mesmerized as the rest of us by the mind-crushing distortion loop that modern TV programming has become.

Foolish Tech Prediction 3

"Nuclear-powered vacuum cleaners will probably be a reality within ten years."

Alex Lewyt, president of Lewyt vacuum company, 1955

Vintage 1950s vacuum cleaner assumptions were that the only thing more certain than the red menace was the inevitability of atomic power. So when New Jersey-based vacuum cleaner honcho Alex Lewyt heralded a tomorrow in which nuclear-powered appliances would suck up dirt in every American household, the news probably caused few eyebrows to rise. Remember, this was the era of radium-impregnated paint for glow-in-the-dark dials. Peaceful radioactivity seemed as safe as asbestos.

Of course, Lewyt's vision has yet to come true, and it likely won't until well after nuclear reactors are enlisted to power all of the terminator robots in our post-SkyNet future.

Foolish Tech Prediction 4

"There is no reason anyone would want a computer in their home."

Ken Olsen, founder of Digital Equipment Corporation, 1977

Digital Equipment Corporation was acquired by Compaq more than a decade ago, but in the 1970s the company was a major force in the world of computing. Apologists argue that DEC president Ken Olsen made this quip before the advent of the PC as we know it, but ready-made personal computers like the MITS Altair had hit the market a couple of years earlier. And within four years of Olsen's remark, the release of the IBM PC had enshrined this prediction in the high-tech hall of shame.

Foolish Tech Prediction 5

"Almost all of the many predictions now being made about 1996 hinge on the Internet's continuing exponential growth. But I predict the Internet will soon go spectacularly supernova and in 1996 catastrophically collapse."

Robert Metcalfe, founder of 3Com, 1995

In addition to being a legendary tech visionary and the man widely credited with having invented Ethernet, Bob Metcalfe was also a columnist for PC World's sister publication InfoWorld. And it was in that column that Metcalfe made what must have been the most regrettable comment of his career; indeed, he even promised to eat his words if his augury turned out to be wrong.

To his credit, Metcalfe made good on that promise in 1999 during his keynote speech at the International World Wide Web Conference, where he blended up a copy of his printed column with some liquid and drank it down before a crowd of onlookers.

Foolish Tech Predition 6

"Apple is already dead."

Nathan Myhrvold, former Microsoft CTO, 1997

To be fair, just about everyone in the computer business thought that Apple was in its death throes when Microsoft CTO Nathan Myhrvold made this comment back in 1997.

Who could have predicted that, a little more than a decade later, that same company would be steadily increasing its share of the PC market while utterly dominating the digital music business and rapidly overtaking the field in the smart phone market?

Foolish Tech Prediction 7

"Two years from now, spam will be solved."

Bill Gates, founder of Microsoft, 2004

By recent estimates, the amount of spam currently glutting up the Net is somewhere around 92 percent of all e-mail messages worldwide. (And it won't do to claim that what he really said was "Two years from now, [Hormel] Spam will be dissolved"-- because the sculptable meat product remains as semisolid as ever.)

4.0 Previous Prediction Theories and Books

4.1 Malthusian Theory

Malthusian Theory is a belief that long term population trends with lack of resources will lead to chaos for the planet. So far it hasn't proved true. (4)

Malthusianism is the idea that population growth is potentially exponential while the growth of the food supply is arithmetical at best. It derives from the political/economic thought of the Reverend Thomas Robert Malthus, as laid out in his 1798 writings.

An Essay on the Principle of Population, Malthus believed there were two types of "checks" that in all times and places kept population growth in line with the growth of the food supply: "preventive checks", such as moral restraints (abstinence, delayed marriage until finances become balanced), and restricting marriage against persons suffering poverty or perceived as defective, and "positive checks", which lead to premature death: disease, starvation, war, resulting in what is called a Malthusian catastrophe.

The catastrophe would return population to a lower, more "sustainable", level. Malthusianism has been linked to a variety of political and social movements, but almost always refers to advocates of population control.

I really think Malthusian Theory has one big drawback—that we don't live in a closed system. The limits of resources might be true if we could only take advantage of what is on Earth, but with access to Space and all the resources it can provide, then there are no limits to support a large population on Earth.

People might even start to migrate off planet to more open and exciting lifestyles.

In fairness to the developers of this theory, when they proposed it in the eighteenth century, they probably could not imagine that there were real opportunities to connect the Earth with Space.

4.2 Future Shock

Alvin Toffler is an Author of several excellent books on future trends. His first book was "Future Shock" published in 1970.

Toffler argued that society is undergoing an enormous structural change, a revolution from an industrial society to a "super-industrial society". This change overwhelms people. He believed the accelerated rate of technological and social change left people disconnected and suffering from "shattering stress and disorientation"—future shocked.

Toffler stated that the majority of social problems are symptoms of future shock. In his discussion of the components of such shock, he popularized the term "*information overload.*"

His analysis of the phenomenon of information overload is continued in his later publications, especially *The Third Wave* and *Powershift*.

Future Predictions By an Engineer and Seer

4.3 The Third Wave

The Third Wave was another book of Toffler's. Its theme was as follows:

In the book Toffler describes three types of societies, based on the concept of 'waves'—each wave pushes the older societies and cultures aside.

- The First Wave is the settled agricultural society which prevailed in much of the world after the Neolithic Revolution, which replaced hunter-gatherer cultures.

- The Second Wave is Industrial Age society. The Second Wave began in Western Europe with the Industrial Revolution, and subsequently spread across the world. Key aspects of Second Wave society are the nuclear family, a factory-type education system and the corporation. Toffler writes:

"The Second Wave Society is industrial and based on mass production, mass distribution, mass consumption, mass education, mass media, mass recreation, mass entertainment, and weapons of mass destruction. You combine those things with standardization, centralization, concentration, and synchronization, and you wind up with a style of organization we call bureaucracy."

- The Third Wave is the post-industrial society. Toffler says that since the late 1950s most countries have been transitioning from a Second Wave society into a Third Wave society. He coined many words to describe it and mentions names invented by others, such as the Information Age.

4.4 Megatrends

In the book "Megatrends" by John Naisbitt in 1982, John predicts that the future generally follow basic grass roots trends.

Here are some of the trends he predicted in his 1982 book:

In 1956-1957, the transition began from an industrial, blue-collar society to an information, clerical, white-collar society. Information is now mass-produced and globally disseminated instantly, yet without selectivity and values.

The second trend is from forced technology to high tech, which balances human response or "high touch" with technology, recognizing that technology cannot solve all problems or do away with the need for responsibility and discipline.

A third trend is the emergence of a global economy replacing national ones. The United States can no longer expect to be the industrial leader but should welcome production sharing and world trade as a contribution to world peace.

Fourth, business management will shift from short-term planning to long-term perspectives, motivated both by concern for the environment and by economic necessity. Banks will have to rethink their function in a world of electronic transfer of funds.

Fifth, America is rapidly decentralizing business, politics, and culture, resulting in a more diverse society, one in which unions, the presidency, and the Congress are obsolete while states and regions are important.

Sixth is the shift from institutional help, provided by government, medical institutions, the school system, and corporations, to self-help through home gardening, hospices, alternative cancer treatment programs, natural childbirth, parental involvement in education, venture capitalism, survivalism, and various self-help groups.

Seventh is the shift from representational to participatory democracy, resulting in a largely nominal two-party system, in the appearance of many nontraditional issues on the ballot, and in activist shareholders and worker involvement in management.

Eighth is the related shift from hierarchies to networks as the pyramid structure collapses. This shift and the previously mentioned one indicate that the facilitator and the empowerer of others will be rewarded rather than the autocrat.

The ninth shift is from North to South, specifically from Northeast to Southwest and Florida, resulting in the emergence of three dominant states (California, Texas, Florida) and ten new cities of rapidly expanding possibilities.

Tenth is the shift from either/or to multiple options, leading to more roles for women, flextime in the workplace, various arts, specialty foods, cable television, and religious variety.

Clearly this book provides important analysis for diverse aspects of American society.

I do agree that long term trends are key to predicting the future, but as I emphasize in this book, you also have to try to take into account potential paradigm changes in doing accurate predictions of the future.

5.0 Science Fiction Predictions

Science Fiction writers make many predictions about the future. Many are fantasy, and many are wrong. However, a few writers have made eerily canny predictions about the future. Did they have prophetic abilities? Here are four famous writers and their predictions. There is also more discussion about writer's predictions in general at the end of the chapter.

5.1 Jules Verne

Jules Verne has an amazing record of predicting developments of technology and events with a nineteenth century viewpoint and having predicted many of the developments in the twentieth century.

In "20,000 Leagues Under the Sea" he predicted a submarine run by the "power of the sun" which could ride underwater indefinitely through the seas. This is quite an amazing statement made by somebody who had no knowledge of Nuclear Power which wasn't even conceived of until the 20th century. Even if you interpret his prediction as talking about solar generated power, solar cells didn't exist and electricity was not in common use at that time.

In his book "From Earth to the Moon" he predicted that three persons would be launched from Florida to the Moon. This was written in 1865 which makes it even more incredible. He said a lot of things about this trip which in retrospect sound more like clairvoyant predictions than just a science fiction story.

But probably his most prophetic work isn't known by most people and was written in 1863 but lost and not published for over one hundred years in 1996. Its title is "Paris in the Twentieth Century" and it has some remarkable predictions.

The book's description of the technology of 1960 was in some ways remarkably close to actual 1960s technology.

The book described in detail advances such as cars powered by internal combustion engines ("gas-cabs") together with the necessary supporting infrastructure such as gas stations and paved asphalt roads, elevated and underground passenger train systems and high-speed trains powered by magnetism and compressed air, skyscrapers, electric lights that illuminate entire cities at night, fax machines ("picture-telegraphs"), elevators, primitive computers which can send messages to each other as part of a network somewhat resembling the Internet (described as sophisticated electrically powered mechanical calculators which can send information to each other across vast distances), the utilization of wind power, automated security systems, the electric chair, and remotely-controlled weapons systems, as well as weapons destructive enough to make war unthinkable.

The book also predicts the growth of suburbs and mass-produced higher education (the opening scene has Dufrénoy attending a mass graduation of 250,000 students), department stores, and massive hotels.

A version of feminism has also arisen in society, with women moving into the workplace and a rise in illegitimate births. It also makes accurate predictions of 20th-century music, predicting the rise of electronic music, and describes a musical instrument similar to a synthesizer, and the replacement of classical music performances with a recorded music industry.

In addition, it predicts that the entertainment industry would be dominated by lewd stage plays, often involving nudity and sexually explicit scenes.

All of this makes one wonder if Jules Verne ever visited the twentieth century.

5.2 Arthur C. Clarke

Arthur C. Clarke was a famous science fiction Author from the mid to later twentieth century. Here are some of the amazing things he predicted:

Communications Satellites

(5) *In a 1964 BBC Special entitled Horizon, Clarke enumerated his personal predictions for the world of 2000 in honor of the 1964 New York World's Fair. In the special he stated that satellites would "make possible a world where we can be in instant contact with each other, wherever we may be" and for those of us living in the present, Clarke had no idea how correct he really was.*

Strangely enough, Clarke's predictions for a global telecommunications network were spawned in 1945 by a 28-year old Clarke in an article that he wrote for Wireless World entitled "Extra-Terrestrial Relays: Can Rocket Stations Give World-Wide Radio Coverage?", regarding the potential uses of geostationary satellites. Clarke theorized that: rocket which achieved a sufficiently great speed in flight outside the earth's atmosphere would never return."

Once the rocket reached orbital velocity, it would become "an artificial satellite, circling the world forever with no expenditure of power — a second moon, in fact." Bear in mind that Clarke's paper was written well over a decade before the former U.S.S.R. sent Sputnik into orbit. Today, telecommunications satellites in geostationary orbits have made the wireless information age possible and it was theorized and predicted in its entirety by Clarke decades in advance.

The Internet

As mentioned previously, Clarke's mother was a radio operator and in this light it becomes clear why Clarke was so intuitive regarding wireless communications, but his theories extended beyond just communications, but also to the internet as a whole.

Clarke's article on geosynchronous satellite networks helped spur interest in the subject, and NASA actually began collaborating with Howard Hughes on the Telstar project in the early sixties (which ultimately set the path for everything from satellite television broadcasts to Hughes Net internet plans as we know them today). As the technology was being developed, Clarke became a mainstay on televisions shows, and when speaking to an Australian news program in 1974 he told the host that by the turn of the century, people would be able to access "all the information needed for everyday life: bank statements, theater reservations, all the information you need over the course of living in a complex modern society." For comparison, online banking services weren't popularized until the late 1990's and by 2000 80% of banks offered online banking, showing that Clarke accurately predicted the rise of the internet, and even further, online banking almost to the year.

Personal Computers

Clarke understood how quickly computers were advancing in the 1960's and 70's and when he made his predictions about the internet and future society, he also described much smaller computer consoles that would allow users to access infinite information.

But the most salient issue he pointed out, at least in this author's opinion, was that Clarke added that people would "take it as much for granted as we take the telephone." Clarke envisioned a world where people would use a console at home to communicate with a computer hub, of one kind or another, that would relay pertinent information back to them.

Clarke had correctly anticipated that ultimately, home computers would enable humanity to do everything from checking their bank accounts to retrieving theater reservations. Clarke had also made allusions to this concept of a connected web that the whole of civilization was tapped into in his book "Profiles of the Future: An Inquiry into the Limits of the Possible", when he talks about generations far into the future as "our remote descendants as living in isolated cells, scarcely ever leaving them, but being able to establish instant TV contact with anyone, anywhere else on Earth."

The iPad

Following Samsung's entrance into the touchscreen and tablet arena, Apple furiously pursued their alleged copyright infringements as Steve Jobs had submitted dozens of patents involving the iPhone's technology. In a desperate move, Samsung cited the movie 2001: A Space Odyssey's "Newspads" as proof that the design was not originally Apple's. Clarke's "Newspads" are practically identical to the popular Apple devices and smartphones and tablets have become fixtures in practically the exact same way they were depicted in the film. There was a passage from the 2001 novel that may even cause you to look at iPads in a more whimsical way: "Here he was, far out in space, speeding away from Earth at thousands of miles an hour, yet in a few milliseconds he could see the headlines of any newspaper he pleased. (That very word "newspaper," of course, was an anachronistic hangover into the age of electronics)."
"Men will no longer commute, they will communicate"

Telecommuting

Clarke made several outlandish and inaccurate predictions including bio-engineered ape servants and the dissolution of cities, but the mechanism that Clarke believed would hasten the fall of urban areas did come true: telecommuting. In his 1964 World's Fair television special, he predicted that by 2000, "men will no longer commute, they will communicate" and that a person could "conduct his business from Tahiti or Bali just as well as he could from London". Clarke said that telecommuting would ultimately prove to be a "wonderful thing," if only in that it meant that people "won't have to be stuck in cities," and they'd be able to live "out in the country" or wherever they want. Anyone in international business will attest that telecommuting and videoconferencing has saved countless hours of travel and agony as well as millions of dollars and that it is almost exactly as Clarke described.

5.3 H.G. Wells

H. G. Wells was an Author around the turn of the twentieth century whose most famous book was "The Time Machine". He didn't have as many predictive hits as Arthur C. Clarke, but he exposed readers to some amazing futuristic concepts.
There is a crater named for him on the far side of the moon because of his book "The First Men in the Moon".

He also wrote a very popular book called "The Outline of History" about the different eras in human history. He was a very influential writer on the then new genre of science fiction. And he made some predictions-many of which came true:

(6) *Phones, Email, and Television*

In Men Like Gods (1923), Wells invites readers to a futuristic utopia that's essentially Earth after thousands of years of progress. In this alternate reality, people communicate exclusively with wireless systems that employ a kind of co-mingling of voicemail and email-like properties.

"For in Utopia, except by previous arrangement, people do not talk together on the telephone," he writes. "A message is sent to the station of the district in which the recipient is known to be, and there it waits until he chooses to tap his accumulated messages. And any that one wishes to repeat can be repeated. Then he talks back to the senders and dispatches any other messages he wishes. The transmission is wireless."

Wells also imagined forms of future entertainment. In When the Sleeper Wakes (1899), the protagonist rouses from two centuries of slumber to a dystopian London in which citizens use wondrous forms of technology like the audio book, airplane and television—yet suffer systematic oppression and social injustice.

Genetic Engineering

Visitors to The Island of Dr. Moreau (1896) were confronted with a menagerie of bizarre creatures including Leopard-Man and

Fox-Bear Witch, created by the titular madman doctor in human-animal hybrid experiments that may presage the age of genetic engineering.

Though Moreau created his Frankenbeasts through more crude techniques, like surgical transplants and blood transfusions, the theme of humans playing God by tinkering with nature has become a reality. Scientists are working towards the day when animal organs could serve as long-term transplants for human patients, though today human immune systems still ultimately reject such efforts. And controversial experiments known as chimera studies create human-animal hybrids by adding human stem cells to animal embryos.

Notably, the human-animal hybrids Moreau creates eventually do the doctor in, and that ending echoes another common Wells theme. "It's often a warning about the consequences of technology, in particular when you don't think them through properly," explains James.

Lasers and Directed Energy Weapons

Martians in The War of the Worlds (1898) unleash what Wells called a Heat-Ray, a super weapon capable of incinerating helpless humans with a noiseless flash of light. It would be more than six decades before Theodore Maiman fired up the first operational laser at California's Hughes Research Laboratory on May 16, 1960, but military thinkers had been hoping to weaponize the conceptual laser even before it was even proven practical.

Wells's description isn't accurate enough to build a working laser, but it resembles both that device and other "directed energy" weapons, such as those using microwaves, electromagnetic radiation, and radio or sound waves, which the United States and other militaries have developed in recent years.

"Many think that in some way [the Martians] are able to generate an intense heat in a chamber of practically absolute non-

conductivity. This intense heat they project in a parallel beam against any object they choose, by means of a polished parabolic mirror of unknown composition, much as the parabolic mirror of a lighthouse projects a beam of light," Wells wrote.

Typically, Wells was more interested in what the effects of his future ideas might be, rather than working out the technical details, James stresses.

"He'll kind of take one element of scientific understanding of the world and tweak it. So in The Time Machine, if you think of time as the fourth dimension, what if you could travel in time as freely as in the other three? Or, in The First Men in the Moon, what if you could make a material [Wells called it Cavorite] as impervious to gravity as other materials are impervious to heat? You just take that one thing, and see what follows from it," James explains.

(Today's leading science fiction authors still use this technique while at work shaping the future of tomorrow. In fact, some companies commission "design fiction" to see how innovative ideas might work if they become fact in the future. "There is nothing weird about a company doing this—commissioning a story about people using a technology to decide if the technology is worth following through on," says novelist Cory Doctorow, whose clients have included Disney and Tesco. "It's like an architect creating a virtual fly-through of a building.")

Atomic Bombs & Nuclear Proliferation

Wells reveled in the potential benefits of technology but also feared their dark side. "H.G. Wells was probably the writer who saw most clearly in the early 20th century the possibility of total war," says Eleanor Courtemanche of the University of Illinois at Urbana-Champaign (A new physical and online exhibition there shows off an extensive Wells collection.)

Wells recognized the world-changing destructive power that might be harnessed by splitting the atom. The atomic bombs he introduces in The World Set Free (1913) fuel a war so devastating that its survivors are moved to create a unified world government to avoid future conflicts.

Wells's bombs differed from those actually developed by scientists with the Manhattan Project. They exploded continually, for days, weeks or months depending upon their size, as the elements in them furiously radiated energy during their degeneration and in the process created mini-volcanoes of death and destruction.

Wells also clearly saw the dangers of nuclear proliferation, and the doomsday scenarios that might arise both when nations were capable of "mutually assured destruction" and when non-state actors or terrorists got into the fray.

"Destruction was becoming so facile that any little body of malcontents could use it; it was revolutionizing the problems of police and internal rule. Before the last war began it was a matter of common knowledge that a man could carry about in a handbag an amount of latent energy sufficient to wreck half a city," he wrote.

Future Predictions By an Engineer and Seer

5.4 Robert Heinlein

Robert Anson Heinlein (July 7, 1907 – May 8, 1988) was an American science fiction writer. Often called the "dean of science fiction writers", his sometimes controversial works continue to have an influential effect on the genre, and on modern culture more generally.

His character "Lazarus Long" appeared in many of his books as a man with extreme longevity who ended up living thousands of years by the last book he appeared in.

He also wrote "The Moon is a Harsh Mistress" about the first colony on the moon and a computer which ran it and had an artificial intelligence personality.

In 1949, science fiction author Robert Heinlein compiled a list of predictions for the year 2000 that were eventually published in February of 1952, in Galaxy magazine. Some were good and some not so good. The list reads as follows:

1. Interplanetary travel is waiting at your front door -- C.O.D. It's yours when you pay for it.

2. Contraception and control of disease is revising relations between the sexes to an extent that will change our entire social and economic structure.

3. The most important military fact of this century is that there is no way to repel an attack from outer space.

4. It is utterly impossible that the United States will start a "preventive war." We will fight when attacked, either directly or in a territory we have guaranteed to defend.

5. In fifteen years the housing shortage will be solved by a "breakthrough" into new technologies which will make every house now standing as obsolete as privies.

6. We'll all be getting a little hungry by and by.

7. The cult of the phony in art will disappear. So-called "modern art" will be discussed only by psychiatrists.

8. Freud will be classed as a pre-scientific, intuitive pioneer and psychoanalysis will be replaced by a growing, changing "operational psychology" based on measurement and prediction.

9. Cancer, the common cold, and tooth decay will all be conquered; the revolutionary new problem in medical research will be to accomplish "regeneration," i.e., to enable a man to grow a new leg, rather than fit him with an artificial limb.

10. By the end of this century mankind will have explored this solar system, and the first ship intended to reach the nearest star will be a-building.

11. Your personal telephone will be small enough to carry in your handbag. Your house telephone will record messages, answer simple inquiries, and transmit vision.

12. Intelligent life will be found on Mars.

13. A thousand miles an hour at a cent a mile will be commonplace; short hauls will be made in evacuated subways at extreme speed.

14. A major objective of applied physics will be to control gravity.

15. We will not achieve a "World State" in the predictable future. Nevertheless, Communism will vanish from this planet.

16. Increasing mobility will disenfranchise a majority of the population. About 1990 a constitutional amendment will do away with state lines while retaining the semblance.

17. All aircraft will be controlled by a giant radar net run on a continent-wide basis by a multiple electronic "brains."

18. Fish and yeast will become our principal sources of proteins. Beef will be a luxury; lamb and mutton will disappear.

19. Mankind will not destroy itself, nor will "Civilization" be destroyed.

Future Predictions By an Engineer and Seer

5.5 Misc Science Fiction Writers

Science Fiction writers in general project many different types of future scenarios.

Some scenarios turn out to be true, and some just wrong. Science Fiction posits so many different possibilities that by chance some will turn out to be true.

However, it is true that some writers are more accurate predicting our future than others. Arthur C. Clarke being a good example.

When I read a science fiction story (and I've read thousands of sci fi books) I always look to see if the Author has followed some train of logical progression to get to his scenario or is it just a fantasy.

A good example are faster than light drives for spaceships. Many different types are proposed to support different space opera environments.

Many of them just claim the ability to access "hyperspace" or some type of "wormholes". While some of these may end up being true I like those stories which try to deal realistically with space drives.

One of my favorites in this regard was the book by Astronaut Buzz Aldrin and Author John Barnes which involves use of lasers to boost a spaceship for a journey to Alpha Centauri. Its title is "Encounter with Tiber".

6.0 The Singularity

I personally think the idea of "The Singularity" is garbage. Having been a software developer and taken many courses on computer design I think that there is no chance of an artificial intelligence becoming "Self Conscious". This is since we humans have a spirit which is the core of our being and lives outside of space and time.

We are not just bags of bone with meat where our intelligence is a result of chemical interactions between our cells.

The idea of "The Singularity" came into being as follows:

The idea was formally described as the "Singularity" in 1993 by Vernor Vinge, a computer scientist and science fiction writer, who posited that accelerating technological change would inevitably lead to machine intelligence that would match and then surpass human intelligence. In his original essay, Dr. Vinge suggested that the point in time at which machines attained superhuman intelligence would happen sometime between 2005 and 2030.

Ray Kurzweil, an artificial intelligence researcher, extended the idea in his 2006 book "The Singularity Is Near: When Humans Transcend Biology," where he argues that machines will outstrip human capabilities in 2045. The idea was popularized in movies such as "Transcendence" and "Her."

Recently several well-known technologists, businessmen, and scientists, including Stephen Hawking, Elon Musk and Bill Gates, have issued warnings about runaway technological progress leading to super intelligent machines that might not be favorably disposed to humanity.

What has not been shown, however, is scientific evidence for such an event. Indeed, the idea has been treated more skeptically by neuroscientists and a vast majority of artificial intelligence researchers.

A Week of Misconceptions

We're using the first week of April as an opportunity to debunk some of the misconceptions about health and science that circulate all year round.

For starters, biologists acknowledge that the basic mechanisms for biological intelligence are still not completely understood, and as a result there is not a good model of human intelligence for computers to simulate.

Indeed, the field of artificial intelligence has a long history of over-promising and under-delivering. John McCarthy, the mathematician and computer scientist who coined the term artificial intelligence, told his Pentagon funders in the early 1960s that building a machine with human levels of intelligence would take just a decade. Even earlier, in 1958 The New York Times reported that the Navy was planning to build a "thinking machine" based on the neural network research of the psychologist Frank Rosenblatt. The article forecast that it would take about a year to build the machine and cost about $100,000.

The notion of the Singularity is predicated on Moore's Law, the 1965 observation by the Intel co-founder Gordon Moore, that the number of transistors that can be etched onto a sliver of silicon doubles at roughly two year intervals. This has fostered the notion of exponential change, in which technology advances slowly at first and then with increasing rapidity with each succeeding technological generation.

At this stage Moore's Law seems to be on the verge of stalling. Transistors will soon reach fundamental physical limits when they are made from just handfuls of atoms. It's further evidence that there will be no quick path to thinking machines.

Future Predictions By an Engineer and Seer

7.0 Secret Science and Knowledge

If I'm going to be totally honest and not hold anything back from my readers then I also have to discuss the claims UFOlogists make that the United States government is holding back secret technologies and the possibility of cooperation with aliens.

In my book "Aliens and Secret Technology- A Theory of the Hidden Truth" I discuss a lot of research and analysis I did of whether there is secret anti-gravity technology and even possibly use of power from "Zero Point Energy".

While I don't have confirmed information which I know for a fact, I believe there is good evidence that the United States government is withholding anti-gravity technology from the public.

Potential alien technology which is hundreds if not thousands of years in advance of our own is also discussed in that book.

The implications are clear. This type of technology when revealed to the public will create a huge paradigm shift which will negate most predictions about our technology and society in the future based on trends analysis.

How then to handle this potential anomalous information?

Since if this turns out to be true, it will have a huge effect on my predictions—I've decided to project that Anti-Gravity is the most surprising technology development which will become commercialized going forward in time.

This means I have it on my list of potential paradigm shifts which will be discussed in more detail later in this book.

Future Predictions By an Engineer and Seer

8.0 Types of Predictions

Predictions can be of several types:

8.1 Predictions based on existing trends

This is the most common type of prediction because it's the easiest. Just project out your trend to its extreme and there is the future.

This has held true for some trends. One of the best is the "Moore's Law" (7) which was propounded by Gordon Moore one of the founders of Intel. In 1975 he projected that microprocessors would keep doubling the number of processors on a chip every two years. He only projected this trend for another decade.

Turns out he was way overly conservative. We can see from the graph below that this trend has now continued for over forty years and is still going on:

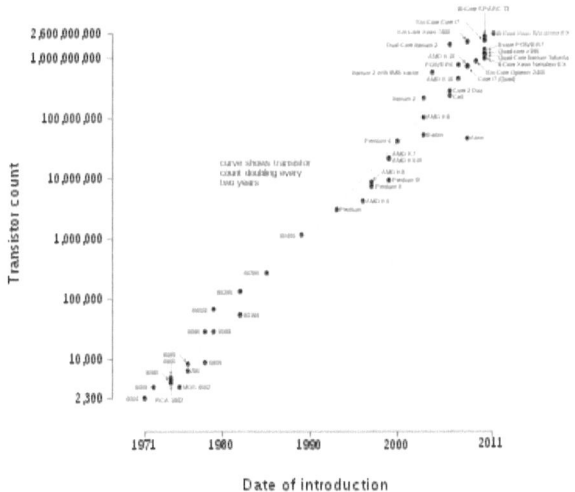

Microprocessor Transistor Counts 1971-2011 & Moore's Law

Now, this trend can't go on forever since there will eventually be size limits reached which are based on the limits of physics, but so far so good.

This example shows how even well-known trends can be misunderstood in terms of how long they will last and their overall impact on the world.

8.2 Potential Paradigm shifts

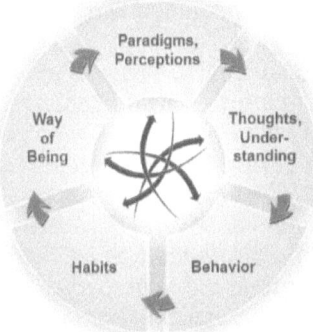

A paradigm shift is very difficult if not impossible to predict. By definition a paradigm shift is some development which changes society so dramatically that nobody is prepared for the results.

The development of the internet is a good example:

The history of the Internet begins with the development of electronic computers in the 1950s. Initial concepts of wide area networking originated in several computer science laboratories in the United States, United Kingdom, and France.

The US Department of Defense awarded contracts as early as the 1960s, including for the development of the ARPANET project, directed by Robert Taylor and managed by Lawrence Roberts. The first message was sent over the ARPANET in 1969 from computer science Professor Leonard Kleinrock's laboratory at University of California, Los Angeles (UCLA) to the second network node at Stanford Research Institute (SRI).

Packet switching networks such as the NPL network, ARPANET, Tymnet, Merit Network, CYCLADES, and Telenet, were developed in the late 1960s and early 1970s using a variety of communications protocols. Donald Davies first demonstrated packet switching in 1967 at the National Physics Laboratory (NPL) in the UK, which became a testbed for UK research for almost two decades. The ARPANET project led to the development of protocols for internetworking, in which multiple separate networks could be joined into a network of networks.

The Internet protocol suite (TCP/IP) was developed by Robert E. Kahn and Vint Cerf in the 1970s and became the standard networking protocol on the ARPANET, incorporating concepts from the French CYCLADES project directed by Louis Pouzin. In the early 1980s the NSF funded the establishment for national supercomputing centers at several universities, and provided interconnectivity in 1986 with the NSFNET project, which also created network access to the supercomputer sites in the United States from research and education organizations. Commercial Internet service providers (ISPs) began to emerge in the very late 1980s. The ARPANET was decommissioned in 1990. Limited private connections to parts of the Internet by officially commercial entities emerged in several American cities by late 1989 and 1990, and the NSFNET was decommissioned in 1995, removing the last restrictions on the use of the Internet to carry commercial traffic.

In the 1980s, research at CERN in Switzerland by British computer scientist Tim Berners-Lee resulted in the World Wide Web, linking hypertext documents into an information system, accessible from any node on the network. Since the mid-1990s, the Internet has had a revolutionary impact on culture, commerce, and technology, including the rise of near-instant communication by electronic mail, instant messaging, voice over Internet Protocol (VoIP) telephone calls, two-way interactive video calls, and the World Wide Web with its discussion forums, blogs, social networking, and online shopping sites.

The result of the incredible internet and World Wide Web is still shaking our world. And if you add Smartphones into the mix, we have an extended paradigm shift of Smartphones able to do things we never did before with the Internet and new technologies like GPS.

Think about how the internet, World Wide Web, smartphones, and GPS have all come together to create an incredible worldwide paradigm shift in the way people live their lives.

Here are just a few of the dramatic changes caused to civilization by the integration of these technologies:

1) Retail stores are closing as more and more people do most of their shopping online.
2) Everyone has cell phones which can surf the web and use many web services wherever they are.
3) Many, many services are offered entirely over the web instead of in person. This even includes medical diagnosis by Doctors in some remote places.
4) Movies are streamed over the web to individuals and homes which has reduced audiences seeing movies at the theater and the video rental business died.
5) The music industry was destroyed and now most people get their music from individual music downloads on the web.
6) Fisherman in poor African villages use their phones and web connections to find where to take their fish for the best market prices.
7) People use their cell phone mapping to find how to get everywhere. Something impossible when I grew up when we used to use paper maps. (Who under 25 can even read a paper map?)
8) Order transportation services using your Smartphone like Uber and Lyft.

What incredible changes the Internet, Smartphones, and related technologies have initiated in our lives. I grew up before the internet and cell phones existed so I know how dramatic the changes have been.

I remember in about 2011 when I was walking at night with my son in Philadelphia looking for a restaurant. We were using my IPhone's mapping capabilities. I saw lots of other people doing the same thing. This is when I became aware that the world had changed dramatically.

So how do we know when other major paradigm shifts are going to occur? This is difficult because by its nature a paradigm shift is a result of the unexpected. All we can do is look for what might become a new paradigm shift and keep track of the changes that it causes.

So can we predict paradigm shifts? Not always, but there are leading technology indicators which help. One example is the development of smart elevators which are electromagnetically controlled and can move horizontally as well as vertically. This will cause tremendous changes to future architecture in buildings and cities. The details are covered later in this book.

My point is that we can't always catch a coming paradigm change but we can make the effort and be partially successful.

8.3 Predictions based on Intuition & Prophecy

Predictions made based on Intuition are a mystery to most and considered ridiculous by the general population too.

Fortunately for me, I've experienced many premonitions in my life and have some idea of how and why they are real.
I related all of this in my book "Prophecy: A History and How to Guide" in 2009.

Here are a few things I learned from my experiences:

a) That real prophecy is tied to our Spirit which we are connected to and which exists outside of space and time. When we clear our minds of emotions and thoughts through activities like meditation, it enhances our predictive intuitional capabilities. It's like a fish tank with a light bulb in it. By stopping the swirling water the sand and dirt falls to the bottom and lets the light shine out.

b) That the future is not fixed and is a set of probabilities. Some probabilities are much stronger than others but all can be changed through force of will. Changing an event like the 911 attacks might be beyond the ability of one person to change, but I have experienced changing taking an airplane flight so I didn't die when the plane crashed killing all aboard.

Here is an example of an intuition which saved my life:

During early August of 1998, my wife and I decided to send her and our kids to visit her mother in Barcelona, Spain.

I was going to buy a ticket separately, and meet them there during early September.

When I started to call the travel agent to book my ticket I had a terrible feeling of fear about taking the flight.

I tried two other times to book the ticket during the week for a September 2nd departure, and each time I got the same strong feelings of fear and death.

I have always prayed and tried to guard myself mentally to avoid disasters, so finally I took the warning seriously and decided not to go at all.

This was very difficult to do since I really wanted to see my wife and kids, and this meant I would be home alone for a month.

Work wasn't an excuse either, since I wasn't doing any really heavy contract work at the time and could easily have taken the time off.

I called my wife and told her my decision, and she was surprised, but agreed for me to follow my instincts.

On September 2nd the Swissair disaster occurred on a plane leaving Kennedy airport in New York, which crashed in Newfoundland Canada with all lives lost.

I would not have originally been booked on that flight, but could have easily ended up on it since I was due to fly through Kennedy airport, and any delay might have caused me to switch planes.

I will never know for sure, but this was a very strong warning.

I should also mention that for several years before this event I had strong feelings that I would be killed in the near future. After this happened those feelings ended.

*****

I have also read a few books by intuitives who attempted to predict the future. One I recommend is by Joseph McMoneagle who wrote "*The Ultimate Time Machine: A Remote Viewer's Perception of Time and Predictions for the New Millennium*"

Future Predictions By an Engineer and Seer

It was an effort to explore the future through remote viewing entirely through clairvoyance. His book *came out in 1998*. His predictions show some very interesting and possibly likely futures over the next 100 years. Here are some of his predictions from that book using remote viewing:

The author foresees a huge expansion in the legal definition of marriage between people, a major trend towards androgyny in clothing and hairstyles, the widespread use of nonpermanent tattoos and the advent of cosmetic scents for both sexes based on pheromones-all designed to evoke certain desires in people within range! Male pattern baldness will be eradicated by ingesting pills, assisted suicide will become a lawful choice, and teenagers will be wearing tracking devices subcutaneously by 2020, for purposes of law enforcement. A new metaphysical religion bridging science and spirituality will emerge; the next pope will ascend to the Chair of St. Peter in late 1999 (but Christianity's overall adherents will decline significantly by 2025); and additional Dead Sea Scrolls will be located between 2011-13.

In the Economics section, by way of a small sampling, McMoneagle sees only four major banks in America by 2030; a common currency for North America by 2040; serious world food production problems by 2039; manned manufacturing companies on the Moon by 2055 and automated manufacturing plants on Mars and two other moons by 2075; a major stock market drop in late 2006, caused by a war in the Middle East; two new car engine types powered, respectively, by on-board, water-separated hydrogen and a new, exotic electrochemistry; bicycles-only transportation restrictions in four American cities by 2025, the business use of Russian submarines for container transport; and the actual teleportation of small physical objects by 2050.

The author views with concern many developments that will occur regarding the Environment. McMoneagle expects air quality to degrade seriously, giving rise to a national epidemic of allergies in children and widespread air-scrubbing in buildings. Major tree-planting efforts will occur globally beginning in China around 2015 as a way of re-"greening" the planet, but not having their full effect until 2080-90. "Designer animals," the products of creative genetic manufacturing, will emerge from 2030-50, originally as pets and then as farm animals, leading to new meat delicacies. (People themselves will be able to shop for desired genetic characteristics in their offspring in 2050!)

For those expecting serious physical earth changes in the offing, McMoneagle's predictions do not disappoint. A continuing rise in average tides will threaten coastal cities worldwide by 2025, according to the author, with Key West and some cities in Europe and the Far East being abandoned around 2045. Underground cities, cities built on platforms, and miniature floating cities will begin to appear during the first 35 years of the 21st century. A new string of Pacific islands will begin to form between Japan and Hawaii after 2041, and the mining exploration of Antarctica will begin in 2020.

Killer earthquakes will occur around the globe, with Los Angeles expected to be hit sometime in 2013-15 and San Francisco during 2022-23, both among the numerous other locales that McMoneagle tabulates in his book for easy reference. A quake occurring in upstate New York in 2050 will be the worst natural disaster in American history. Hurricane activity will also become more frequent, severe, and deadly, with people abandoning island living in the Bahamas and Virgin Islands by 2025 as a result. An asteroid passing Earth in 2016 will have an electromagnetic effect on the Earth's surface, and the existence of a tenth planet in our solar system will be verified by telescope in 2015.

8.4 Taking into Account Economics and Industry

One of the factors which many predictions fail to take into account is economics.

Is the prediction based on cost effectiveness of how people would apply themselves to a new technology, development, or opportunity?

Is the prediction realistic in terms of the money needed to accomplish what is claimed? A good example of this would be Space Colonies. I put a lot of thought and analysis into my book "Designing and Building Space Colonies - A Blueprint for the Future".

One thing which became clear to me was that although lots of space colony proposals might be doable technologically, the billions of dollars needed to build them would take a huge amount of industrial justification and commercial incentives.

So looking at the economic justification for accomplishing certain goals is as important as whether these goals can be accomplished with the technology available.

Most Science Fiction and projections of the future totally ignore economic reality. Much as I love the Star Trek Universe, they never provide any type of background or justification of how that futuristic economy works.

9.0 Mixing Logic and Knowledge with Intuition

So what I'm attempting in this predictive book is to mix knowledge of trends, thoughts about possible paradigm changes, and analysis, with intuitive or Prophetic predictions.

My thinking is to use known trends to form my main predictions and then use my intuition to see if these predictions "Feel Right" in the future.

I do this a lot with short term events to feel if they are "Bright" or "Dark". I'm usually right and it makes life interesting. My intuition has also saved my life numerous times as stated previously.

Hopefully my intuition will give me some insight into when paradigm shifts will occur mixed in with the projection of trends to form better overall predictions.

Paradigm shifts can often be guessed at based on potential confluences of technological developments.

I think that by doing detailed research on trends, an analysis of coming paradigm changes, then adding some intuition about what potential futures make sense—I can develop a more accurate picture of what our real future will be than many other people.

Future Predictions By an Engineer and Seer

Part 2-Major Trends & Predictions

10.0 World Population Growth

Global human population growth amounts to around 83 million annually, or 1.1% per year. The global population has grown from 1 billion in 1800 to 7.6 billion in 2017. It is expected to keep growing, and estimates have put the total population at 8.6 billion by mid-2030, 9.8 billion by mid-2050 and 11.2 billion by 2100. Many nations with rapid population growth have low standards of living, whereas many nations with low rates of population growth have high standards of living. (8) Here is a population growth trend for worldwide population growth projected by the United Nations as of 2017:

Years passed	Year	Population Billion
-	1800	1
127	1927	2
33	1960	3
14	1974	4
13	1987	5
12	1999	6
12	2011	7
12	2023*	8
14	2037*	9
18	2055*	10
33	2088*	11

*World Population Prospects 2017
(United Nations Population Division)

This huge population growth obviously has implications for the quality of life on our planet and the need for resources to support it.

I think there is a good argument that as countries become more advanced and developed technologically, and their populations become more educated, there is a corresponding drop in the average number of babies a family has. So population growth is slowing down worldwide and has a very good chance of stabilizing in the next hundred years or so.

So my belief is that the current developments in infrastructure and food production to support a growing population will continue to keep the worldwide population from starving.

We are also continuing to have a more efficient usage of resources to support ourselves.

A good example is one I cited in an earlier chapter about Fishermen in Africa using Smartphones and the Internet to find the best places to sell their fish which results in less waste and more profits.

11.0 Availability of Resources

The availability of resources for manufacturing and food are a big issue of Malthusian Theory. The idea being the Earth is a closed system and we are close to maxing out resources from Earth. This does not seem to be correct because as some resources become scarce at certain prices, higher prices create the opportunity for newly available resources.

A good example is the availability of oil and gas to power our society. I remember the 1970s when because of the Arab and Israeli war in 1973 the Arab countries imposed and oil embargo on the west.

Also, that we fought a couple of Gulf Wars to maintain access to Middle Eastern Oil Resources.

Prices in 2009 were over $140 per barrel and the many predictions were that we had entered a time of scarcity of oil worldwide. That the United States would be dependent on foreign oil from now on.

Then a major paradigm shift occurred. This was the development of Fracking by United States oil companies. Fracking was all about horizontal drilling technologies and pumping fluids into oil shales to crack it and release petroleum and natural gas.

The result was that oil and gas prices came way down and have stabilized today at $50-$60 per barrel. Higher than pumping oil originally costed, but still a very economical price.

What have the impacts been of this huge paradigm change?

1) Countries who formerly counted on oil for most of their income have had to cut way back and find other sources of revenue. This had led to major changes in the economies of Russia, Saudi Arabia, Venezuela, and others.

2) The United States has become an oil and gas exporter and the totals (as of now in 2018) are going up every month.

3) With new reserves available due to new Fracking technologies, the recoverable oil and gas in the United States are now the largest proven reserves in the world.

Who would have ever guessed that the United States would become the largest producer of oil in the world again in the twenty first century? It is estimated that will happen by the end of 2018.

The world has gone from believing that we have reached and passed "Peak Oil" to the reality that with Fracking the world is nowhere near worries about "Peak Oil" production.

And this shifts balances militarily too. We fought two Gulf Wars over concerns about the availability of Mideast oil supplies. With Fracking, this no longer a concern so it changes the balance of Middle East political and military thinking.

*****

Similar wrong thinking is true about the ability of the world to grow food. As the world population grows, scientists and farmers find new ways to dramatically increase food production through genetic improvements in crops, and other types of farming like hydroponics.

There are also companies who are looking to open up the availability of resources by mining asteroids. While the end results of this development of technologies to support it might be decades in the future, it is clear that Earth's population is in no way constrained for resources the ways that Mathusians originally predicted.

12.0 Technology Development Trends

Additional Technologies being developed will have profound effects on our civilization in the future. In this chapter we will look at trends in these major technologies and where some potential paradigm shifts might occur.

It is amazing how rapidly many of these diverse subjects are developing.

12.1 Artificial Intelligence

Many industrialists and some scientists have a great fear of artificial intelligence. Fortunately, most of them don't really understand software or artificial intelligence systems so I think most of their worries are from ignorance.

Having been a software developer much of my career and keeping current with research on Artificial Intelligence (AI) I'm not very worried.

The most recent practical systems are called Neural Networks where software learns from repeated trying different results in a scenario to find the optimum results. This is a form of machine learning but is not "Intelligence". Instead it's all about complex algorithms which help the software follow a learning curve from repeated experiences input into it.

Another reason I don't worry much about a "Rogue AI" are my experiences and beliefs that we humans are not just mechanical systems, but we have a divine spirit which exists outside of ourselves and outside of space and time. Computers and software don't have this.

So what developments do I see? Here are some possibilities:

1) AI learning systems can help with automated driving. Many systems are already being tested and "learned" information is collected and fed back into these computers. This

technology is not ready to go but we might start seeing more of it in ten to twenty years.
2) Better speech recognition. We are already seeing the results in products like the IPhone's Siri and other devices which recognize our speech to perform tasks or launch other programs.
3) Language translation. As AI learning systems and speech recognition become more effective then auto language translation systems will become more popular.
4) More automated systems like spaceships and manufacturing technologies to take the drudge work from humans.

Why do famous scientists and technologists like Elon Musk, Steven Hawking, or Bill Gates all worry about Artificial Intelligences dominating our lives?

Again, I think it's because they think we are just biological machines not spiritual beings and worry that intelligence can be created much more easily than what the true reality is.

I do not see a huge all of a sudden paradigm shift caused by AI systems in our future. AI systems will continue to take over boring and/or repetitive jobs, but new ones will be created in the process.

12.2 Biology & DNA

DNA analysis and modification technologies have rapidly improved in recent years. Now the development of the "CRISPR" technology has revolutionized the ability of scientists to modify DNA.

CRISPR technology is a simple yet powerful tool for editing genomes. It allows researchers to easily alter DNA sequences and modify gene function. Its many potential applications include correcting genetic defects, treating and preventing the spread of ethical concerns. (9)

In popular usage, "CRISPR" (pronounced "crisper") is shorthand for "CRISPR-Cas9." CRISPRs are specialized stretches of DNA. The protein Cas9 (or "CRISPR-associated") is an enzyme that acts like a pair of molecular scissors, capable of cutting strands of DNA.

CRISPR technology was adapted from the natural defense mechanisms of bacteria and archaea (the domain of single-celled microorganisms). These organisms use CRISPR-derived RNA and various Cas proteins, including Cas9, to foil attacks by viruses and other foreign bodies. They do so primarily by chopping up and destroying the DNA of a foreign invader. When these components are transferred into other, more complex, organisms, it allows for the manipulation of genes, or "editing."

With development of CRISPR the modification of DNA to cure disease and make basic DNA changes to existing biological systems starts to get more and more real.

Considering the speed of advances in DNA analysis and change the potential for a large paradigm shift exists within our lifetimes to cure major diseases and make major modifications to plants and animals—including humanity.

In a later chapter we consider the implications of genetic improvements to human intelligence.

Future Predictions By an Engineer and Seer

12.3 Construction

Building construction continues to advance around the world, with increasingly complex building designs and much higher towers being built.

a) Height

The Jeddah Tower (Jeddah, Saudi Arabia: 3,281 feet, estimated completion 2020) is currently the tallest tower under construction.

Frank Lloyd Wright propose a mile high tower in a book of his in 1957. It seemed outrageous at that time, but the new Jeddah tower will be two thirds of a mile high—so we are approaching this type of science fiction design in reality.

Towers which were once considered fantasy are actually being built. We could build even taller and stronger towers using materials like super strong carbon composites.

Over the next few hundreds of years, we should see towers of two miles, three miles high, and more. It's all a matter of design technologies and materials

b) Horizontal Elevators

An additional exciting technology will be elevators which will get rid of their cables and use electromagnetic technology to also go horizontally.

This means faster elevators can go around slower elevators and even outside their home buildings.

Imagine a city with strands connecting many buildings together at multiple heights. Conceivably you could enter an elevator in one building and end up across town on a different floor in a different building.

How long will it take to develop these electromagnetic elevators?

One company-ThyssenKrupp unveiled a revolutionary multi-directional elevator concept in 2014. The concept is still undergoing testing, but it looks like this technology will start to be deployed in the next few years.

It will probably take decades for this technology to take off in multiple connected buildings since it will require major retrofits to those buildings or entirely new construction.

But, imagine a city which has interconnections at many different levels, and where you can travel anywhere in the city from getting into just one elevator!

As a paradigm change it will cause the designs of our buildings and interconnectivity to change over the new decades and centuries.

12.4 Crypto Currencies

A cryptocurrency (10) (or crypto currency) is a digital asset designed to work as a medium of exchange that uses cryptography to secure its transactions, to control the creation of additional units, and to verify the transfer of assets.

Cryptocurrencies are classified as a subset of digital currencies and are also classified as a subset of alternative currencies and virtual currencies. Cryptocurrencies use decentralized control as opposed to centralized electronic money and central banking systems. The decentralized control of each cryptocurrency works through a blockchain, which is a public transaction database, functioning as a distributed ledger.

Bitcoin, created in 2009, was the first decentralized cryptocurrency. Since then, numerous other cryptocurrencies have been created. These are frequently called altcoins, as a blend of alternative coin.

Crypyto currencies are very new on the world stage so it's very hard to tell what impact they will have on our civilization. They do fill a need though which is that people don't want to have to depend on a central bank for their currencies. This is the first stateless currency every invented.

In 2018, United States Futures markets are even offering trades on the largest crypto currency—Bitcoin. So acceptance is building.

If the world did become dominated by crypto currencies then it would keep governments from inflating their money or just creating money for politicized goals and control. This would be a good thing.

Take the most important currency in the world—The United States dollar. The dollar is not backed up by anything except people's faith in the United States government and its credibility.

How to predict if Crypto Currency will become a major paradigm shift? That's depends on how many people are affected and the impact of these effects on society.

If the current world currency regimes all crashed—similar to what happened in 1929 at the beginning of the Great Depression—which was worldwide—then usage of Crypto Currencies might be the only way to give people confidence in the value of currencies again.

In this situation people may no longer have confidence in any of the major current state currencies and so crypto currencies may become new default for world exchange.

So I'd put the Paradigm Change outlook as moderate in the next 100 years.

12.5 Drone Technology

Flying Drones are a really interesting development in the last twenty years which became mainstream when they started conducting aerial reconnaissance missions for the military.

The availability of GPS has also provided easier ways for drones to navigate over long distances.

The usage of automated drones is growing fast and some future projections of their usage include:

1) Delivery of packages by Drones. This is already being pioneered by Amazon.com
2) Swarms of autonomous Drones for military or other usages. This is like a flock of birds where the rest of the drones all follow one leader. Could be used militarily or for say applying pesticides to farm fields.
3) Drones for refueling manned aircraft. This is already being planned by the United States Navy.
4) Tiny Drones which can be used for military surveillance, by the police, or by someone like a forester doing surveys of a large area of forest

In the future drones will become much smaller and more common in all of our lives.

I predict that this trend will continue to grow for the next fifty to one hundred years.

12.6 Education

Education is changing in many ways today. Distance learning is a reality over the internet. Some companies sell online courses, and more and more training is online. Universities offer their courses online and even degree programs over the internet.

So the question is how much will training move out of the classroom and online? And will it be just Universities or Primary and Secondary schools too?

There are already experimental high schools online for families doing home schooling.

More and more textbooks are offered electronically too so students can read them on IPads or Laptop computers.

Also, instead of rote learning, more students will be judged and tested on creativity since almost all of the facts will be available online.

Schools will gradually move to being present everywhere you have your web connected devices.

This is a gradual paradigm change going on today. At some point in the future a large part of education will be delivered electronically and remotely.

Classrooms will still exist but mainly for the importance of personal interaction and allowing teachers to help students directly and one on one.

I envision that eventually there will be many fewer physical colleges and universities since there will be such excellent courses offered online and through virtual reality.

However, there will still be some of the more prestigious schools with physical student bodies since students and parents will still be excited about learning in person at an exclusive educational institution like Harvard or Stanford.

But many local colleges and lesser known Universities will disappear in favor of high quality online learning offerings.

12.7 Farming

Improved farming will result from several technologies.

Genetic Engineering will continue to improve crop yields and hardiness from diseases. This has been going on since the 1970s and will continue to show benefits to farming.

Hydroponics is also considered a big future growth opportunity for farming since many areas of the world now have limited open arable land.

(11) Hydroponics is the fastest growing sector of agriculture, and it could very well dominate food production in the future. As population increases and arable land declines due to poor land management, people will turn to new technologies like hydroponics and vertical farming to create additional channels of crop production.

Currently, arable land comprises only around 3 percent of the Earth's surface, and the world population is around 6 billion people, resulting in around 1/5 hectare (2,000 square meters) of arable land per capita. By 2050, scientists estimate that the Earth's population will increase to 9.2 billion, while land available for crop and food production will decline. To feed the increasing population, hydroponics will begin replacing traditional agriculture.

To get a glimpse of the future of hydroponics, we need only to examine some of the early adopters of this science. In Tokyo, on the island nation of Japan, land is extremely valuable due to the surging population. To feed the citizens while preserving valuable land mass, the country has turned to hydroponic rice production. The rice is harvested in underground vaults without the use of soil. Because the environment is perfectly controlled, four cycles of harvest can be performed annually, instead of the traditional single harvest.

Hydroponics also has been used successfully in Israel, which has a dry, arid climate. A company called Organitech has been growing crops in 40-foot (12.19-meter) long shipping containers, using hydroponic systems. They grow large quantities of berries, citrus fruits and bananas, all of which couldn't normally be grown in Israel's climate.

The hydroponics techniques produce a yield 1,000 times greater than the same sized area of land could produce annually. Best of all, the process is completely automated, controlled by robots using an assembly line-type system, such as those used in manufacturing plants. The shipping containers are then transported throughout the country.

The future on Earth and in space should see the growth of hydroponics to help provide a steady stream of food to growing populations.

12.8 Electrical Power Generation

Electrical Power generation comes from multiple sources and so has multiple trends to follow.

Solar Cells

Solar Cells keep getting more efficient and increasing in usage as their efficiency and cost effectiveness grows. From the below table you can see the trend of efficiency over time for different solar cell technologies:

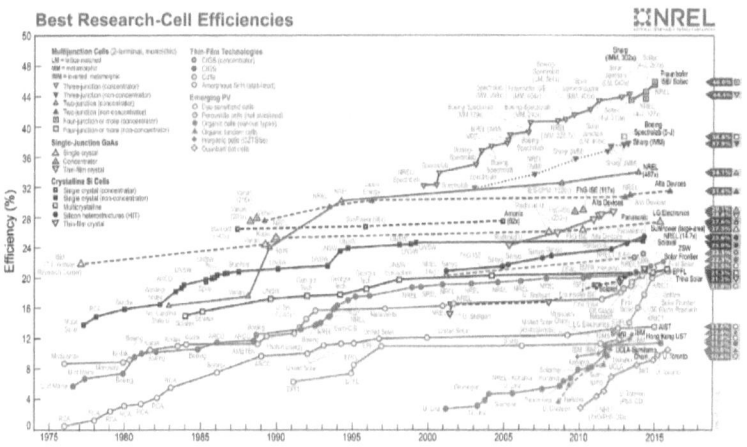

Some solar cells are approaching a 50% efficiency conversion level which would have been considered ridiculous fifty years ago.

It's fair to assume that solar cells will keep penetrating the market for the next few decades as they become more and more cost effective.

However, we are getting closer to theoretical limits for electric conversion efficiency so I don't see radical or paradigm changing improvements in solar cells in the future.

Solar Cells should continue to have reduced costs and will become economically feasible in more and more applications.

Nuclear Fusion

I studied Nuclear Fusion in engineering school in the 1970s. Nuclear Fusion plants producing electricity were supposed to be thirty years away at that time. Today, although advances have been made, the industry still seems thirty years away from production.

Eventually, Nuclear Fusion should go into commercial usage in the next few decades. Electricity costs might drop a lot in the next fifty years due to Nuclear Fusion electricity production finally going online.

Nuclear Fusion doesn't produce radioactive wastes like our current fission technologies and could also be scaled to different sizes so the potential is huge.

Battery Energy Density

Batteries are critical to storing electrical energy as well as powering mobile devices. The development of batteries is about enhancing their energy density so that they can hold more energy per volume.

Today the highest energy density of batteries in large production is for lithium based batteries.

Here is an article on increasing energy density in batteries over the coming years:

(12) *Press releases from research institutions always offer tantalizing new technologies but you know there are still many hurdles along the way before the product will finally hit the shelves. There are of course research teams worldwide working on novel battery technologies because efficient energy storage systems will become increasingly important in the future.*

Currently on offer we have new cell technologies that promise energy density increases of 3 or 15 times that of conventional lithium cells...

At Rice University in the US, researchers are working on a lithium-metal cell whose capacity exceeds the current Li-Ion batteries by a factor of three. The special features of this cell technology are not only its improved capacity but also no dendrite growth in the cell. Dendrites are those pesky whiskers of lithium that grow in the cell over time and eventually cause short circuits. The trick here is to coat a material with a very high surface area of carbon nanotubes with metallic lithium. The result is a safe battery with a capacity of 3.351 Ah / g. The prototype cell still retains 80% of the original capacity after 500 charge cycles.

The NIMS (National Institute for Materials Science) in Japan is currently researching lithium-air batteries. This technology potentially offers a capacity increase over conventional Lithium batteries of a factor of 15! Li-Air batteries have in principle a much higher energy density because a lot of electrode material is dispensed with. If the capacity achieved in the lab (30 mAh / cm²) were to be realized in a commercial product that would be sensational. Again, the electrode material has an enormous surface area thanks to carbon nanotubes. Work is ongoing to produce real practical samples with high energy density and a system to filter impurities from the air.

So we can expect energy density improvements of 3-15 times in the near future. Beyond that a new storage technology would be needed.

When you look at Solar Cell improvements, Nuclear Fusion and Battery Energy density increases it is reasonable to expect some good improvements in the cost effectiveness of electricity production and storage in the rest of the 21st century.

This will lead to more electric cars and trucks and increased cost effectiveness overall in technology using electricity.

Do I see a huge paradigm shift in this area? Not immediately, just generally increasing efficiencies over time.

Power costs in the future may well be lower than those today due to advanced storage technologies and low cost baseline power from Fusion plants.

12.9 Human Longevity

Longevity is a subject I know a lot about since I've written numerous books on Longevity and even developed an online training Longevity Coaching program in the last five years. (See my site at: http://personal-longevity.com)

There are many people who have lived well into their one hundreds and a few into their two hundreds and more.

Here is a case study of LI-Ching-Yung who lived into his two hundreds:

Below is an excerpt of an article from the New York Times:

<u>*The New York Times, Saturday, May 6, 1933*</u>

LI CHING-YUN DEAD; GAVE HIS AGE AS 197

He said "Keep a Quiet heart, Sit Like a Tortoise, Sleep Like a Dog," His advice for a Long Life. Inquiry Put Age at 256.

Reported to have buried 23 wives and had 180 descendants – sold herbs for the first 100 years.

Peiping, May 5 – Li Ching-Yun, a resident of Kaihsien, in the Province of Szechwan, who contended that he was one of the world's oldest men and said he was born in 1736 – which would make him 197 years old – died today.

A Chinese dispatch from Chungking telling of Mr. Li's death said he attributed his longevity to peace of mind and that it was his belief every one could live at least a century by attaining inward calm.

Compared with estimates of Li Ching-Yun's age in previous reports from China, the above dispatch is conservative. In 1930 it was said Professor Wu Chung-Chien, dean of the department of Education in Minkuo University, had found records showing Li was born in 1677 and that Imperial Chinese Government

*congratulated him on his 150th and 200th birthdays.
A correspondent of The New York Times wrote in 1928 that many of the oldest men in Li's neighborhood asserted their grandfathers knew him as boys and that he was then a grown man.*

According to the generally accepted tales told in his province. Li was able to read and write as a child, and by his tenth birthday had traveled in Kansu, Shansi, Tibet, Annam, Siam and Manchuria gathering herbs. For the first hundred years he continued at this occupation. Then he switched to selling herbs gathered by others.

Wu Pei-Fu, the warlord, took Li into his house to learn the secret of living to 250. Another pupil said Li told him to "keep a quiet heart, sit like a tortoise, walk sprightly like a pigeon and sleep like a dog."

According to one version of Li's married life he had buried away twenty-three wives and was living with his twenty-fourth, a woman of '60.' Another account, which in 1928 credited him with 180 living descendants, comprising eleven generations, recorded only fourteen marriages. This second authority said his eyesight was good; also, that the finger nails of his right hand were very long, and "long" for a Chinese might mean longer than any finger nails ever dreamed of in the United States.

One statement of The Times correspondent which probably caused skeptical readers to believe Li was born more recently that 1677, was that "many who have seen him recently declare that his facial appearance is no different from that of persons two centuries his junior."

Here are my predictions for Longevity:

1) We can learn how to improve our longevity by many decades through integrating spirit, mind, and body. This will lead to healthy living well into our one hundreds.
2) Science is developing more knowledge about what things can help our bodies live a long time. One area is telomeres

which are the endcaps of chromosomes. As we learn more we will be able to better protect our cellular telomeres which should lead to slower aging of the body.

3) Life expectancies will keep rising across the world due to healthier lifestyles and improvements from science. This should lead to double today's average life expectations within the next 100 years.

The obvious question is will this improved longevity become a major paradigm shift?

I would say yes—but it will occur over centuries as our overall life expectancies continue to increase.

As I discussed in my Science Fiction Novel "Personal Freedom- Parts 1 & 2" people may start living long enough to make trips in interstellar spaceships which last hundreds of years and still be alive and healthy when they arrive at the new star!

As life expectancies increase people will have more time to enjoy multiple careers and childhood may last decades instead of just until they reach eighteen years old.

With a much longer life people might have several long term marriages and families, get multiple degrees at different times in their lives, and have multiple different careers.

The quality of life will rise dramatically for everyone with vastly longer lives where one can have multiple careers, relationships, and see much more of our world and even outer space in one lifetime.

12.10 Manufacturing

3D printing is itself a major paradigm shift. Previously parts had to be designed in blueprints, then carefully machined and then assembled in many cases with machining limitations taken into account.

Now one can design something on a computer, send the design to a 3D printer and build the object with all of its complexity in the 3D printer quickly.

In fact, more and more, things are being printed which would be too complex or impossible to machine. Examples are rocket engines with internal cooling structures designed and 3D Printed. Some of these designs would be impossible with conventional manufacturing technologies.

3D printing experiments were only started in the 1980s and the technology is just starting to enter industry in a big way.

Also, the materials which can be printed used to be limited to plastics but now metals and even biological materials are being printed.

The growth in the development and acceptance of 3D printing should go on for many decades since there is still a lot of development and innovation going on in this area.

Eventually, most manufacturing will use the 3D approach. Even homes and buildings will be constructed with the 3D process since it is faster and less costly than conventional construction methods.

Some 3D printers should become so inexpensive that people will buy them for their homes (hobbyists already do) to print clothes, dishes, and many household items.

So a revolution has started in Manufacturing which may continue for another hundred years or more as more complex things are built using 3D printing technologies—including houses.

Think about some of the possibilities:

- Building a home using a large 3D printing machine from electronic blueprints with a significant reduction in labor. Instead of it taking months the whole process now takes weeks.
- Use 3D printing to print food. Flavors can be added automatically and you can vary the recipe just by changing the "blueprint recipe".
- Print new organs for medicine from cell structures as the materials. There no longer needs to be a shortage of hearts of kidneys for people who need them.
- Print intricate structures which can't be machined or assembled any other way. This is already being done to print rocket nozzles for SpaceX rockets.

I think 3D printing will continue to affect our lives in a big way for possibly the next few hundred years as this technology matures.

12.11 Medicine

Improvements in Medicine should be looked at from several perspectives.

First, is that DNA and Genetics modifications from technologies like CRISPR will result in curing many genetically caused diseases and might even result in cures to other diseases being incorporated into people's DNA.

(13) Here are some additional expected future improvements in Medicine from Technology:

- Organs created from 3D printing then implanted into the patient
- An arsenal of home diagnostic tools
- Remote diagnostics and treatments by Medical Professionals—even in Space
- Robotic and mechanicals aids to those with impairments or lack of mobility to let them live more normal lives
- Risky Medical procedures made simpler through imaging diagnostics and imaging displays of what is going on inside the body.

Paradigm changes? More gene editing using CRISPR technology can lead to some dramatic cures and modifications to our bodies.

Overall, the future of medicine looks like a continual advancement to cure diseases and other conditions which limit life experiences today.

12.12 Mobile Technologies

Mobile technologies have changed tremendously in the last few decades. In the 1980s we had clunky cell phones like the "Brick" phone.

Then we had flip phones in the 1990s as it became more and more popular.

The IPhone was announced in 2007 and it popularized the idea that everyone should carry a Smartphone which had GPS, and could run many types of apps—and also be a telephone. Now almost everyone has or wants a Smartphone.

As of the beginning of 2017 over fifty percent of people accessed the web using their mobile devices. That is an incredible growth trend.

The Smartphone and cell phones which went before can be considered a paradigm change from what went before. With the inclusion of internet and web access, digital cameras, and GPS availability for all Smartphones- a true revolution in the way people live and work has occurred over the last thirty years— with many societal and social implications.

What else can we expect from Mobile technologies?

1) After researching this quite a bit, there will be some innovations like more capabilities for sports and fitness tracking.
2) There is also the potential for some type of brain interfaces to let thoughts control the actions of mobile devices.
3) Flexible and Wearable Smartphones
4) Augmented reality with more 3D applications.

However, I'm disappointed to say that mobile Smartphones seem to be reaching a level of maturity pretty quickly. By 2020 I'm not sure we will see many more improvements.

So Mobile Phones have already caused one major paradigm change—I don't see another major change from Smartphones on the horizon.

I think augmented 3D reality apps are nice but I don't see them changing the world the way the original smartphone has done.

12.13 Military Technology

Military Technology is always one of the fastest advancing areas of human development, since it directly affects the ability of nations to have more power and control over their foreign policies.

Some advances currently in the works include:

1) Development of Lasers for combat. It is expected that within the next five to ten years the United States will be fielding lasers for combat on ships and in airplanes. This will improve the cost effectiveness and speed of weapons delivery dramatically.
2) Exoskeletons to dramatically improve the carrying capacity and speed of troops on the ground. This will allow them to carry more weapons and move to their objectives much faster. Prototypes are already being tested.
3) Self-Guided Bullets to Increase the Lethality of guns significantly.
4) Increased Sensor Fusion beyond today as illustrated by the F-35 fighters starting to come online
5) More usage of Space as an ultimate "High Ground" and the general weaponization of Space

Although Military technology and Weapons keeping improving continuously I don't see any major paradigm changes unless... if the rumors of secret anti-gravity technology transportation are true....that would be a game changer for not just military, but many aspects of our society.

We do need to discuss Nuclear Weapons and their proliferation. Basic Nuclear weapons technology is almost seventy five years old. Any Nuclear Engineering University student today can design a nuclear weapon. (I know because I was one in the mid-1970s and can state this from my own knowledge.) Most nuclear weapons information can be found on the internet.

The biggest thing limiting nuclear weapons proliferation is the availability of uranium and plutonium, the ability to handle it safely, and the money needed to refine or create it from scratch. As the tools become less expensive to get refined nuclear materials, the availability of "The Bomb" will become even more common around the world.

I don't think the great powers like the United States, Russia, or China will have nuclear wars against each other because they have many safeguards on their weapons and know that having a nuclear exchange with another great power would likely be suicidal. (Mutual Assured Destruction).

But, I have recently read about a new Russian undersea "Doomsday" weapon. It is a submarine drone designed to go for thousands of miles and carrying a 100 megaton warhead. Blow one up off shore of a major city like New York, and the city will cease to exist.

The biggest danger then is from small states—like North Korea who may have unstable leaders and a willingness to use Nuclear Weapons to achieve their aims. These might be major miscalculations on their parts, but they will eventually attack someone with nuclear weapons.

I think the likely scenario is that there will be a limited nuclear exchange initiated by a small unstable governmental power, which will lead to limited nuclear fallout in the areas attacked or from nuclear responses.

In the long run this should not affect human civilization too much since again, I think the exchanges will be limited and the unstable smaller powers will be destroyed as a result.
So I don't think a worldwide nuclear war and high radiation or climate effects will be likely in the future.

However, the military will move more and more into Outer Space as the ultimate high ground. This means more military astronauts, and more weaponization of Space.

Maybe even "Rods From God" or tungsten rods which can be dropped from space to affect military conflicts on the ground. These penetrating rods can have the same power as small nuclear weapons without the radiation dangers.

Long term trends indicate continuing expansion and improvements in military technologies but nothing dramatic is on the horizon.

12.14 Nanotechnology

Nano technology is manipulation of matter on an atomic, molecular, and supramolecular scale. Nanotechnology as defined by size is naturally very broad, including fields of science as diverse as surface science, organic chemistry, molecular biology, semiconductor physics, energy storage, microfabrication, molecular engineering, etc.

The associated research and applications are equally diverse, ranging from extensions of conventional device physics to completely new approaches based upon molecular self-assembly, from developing new materials with dimensions on the nanoscale to direct control of matter on the atomic scale. Nanotechnology (nanotech) is still in the early developmental stages and we don't know how long it will take to implement it seriously in different industries.

When feasible though, think about nanotech being used in the human body to get rid of disease, to repair our bodies from injuries by rebuilding flesh and organs too.

Nanotech could also have major effects on materials, giving them new properties, allowing them to change shape, and change their properties.

Nanotech could also be used a weapon. The "Grey Goo" scenario is that nanotech could disassemble everything in the world into its sub components on an atomic level and just leave the surface of the Earth in a single "Grey Goo" substance.

Another more positive scenario has to do with nanotech machines creating their own duplicates to affect things on a macroscopic level to create for instance a new colony in space from nanotech assemblers. This might be in the far future, but huge macroscopic projects are possible through Nanotech.

This could also lead to home "molecular assemblers" which could assemble food for meals, new clothes, etc. from electronic designs.

So how long will it take for nanotech to affect our world?
One could argue that this has already been happening with the developed of increasingly smaller integrated circuits.

So the development and widening use of nanotech seems to be something which is happening over many decades.

Think about some far future applications of nanotechnology where you draw the design on intelligent materials and the materials are smart enough to create the design with specific properties from just a blueprint.

The far out ideas of the applications for nanotech will probably happen, but some of these applications might take centuries to develop.

So nanotech offers multiple paradigm changes for many industries, but it seems to be multi-generational in its effects on our descendants and many future generations may still be feeling the effects of the development of this technology.

12.15 Quantum Technologies

There are at least a couple of ways our understanding of the Quantum world is leading to new technologies:

A) Quantum Computing

Quantum computing is computing using quantum-mechanical phenomena, such as superposition and entanglement. A quantum computer is a device that performs quantum computing. They are different from binary digital electronic computers based on transistors. Whereas common digital computing requires that the data be encoded into binary digits (bits), each of which is always in one of two definite states (0 or 1), quantum computation uses quantum bits, which can be in superpositions of states.

Large-scale quantum computers would theoretically be able to solve certain problems much more quickly than any classical computers that use even the best currently known algorithms, like integer factorization using Shor's algorithm or the simulation of quantum many-body systems. There exist quantum algorithms, such as Simon's algorithm, that run faster than any possible probabilistic classical algorithm.

So Quantum computing should lead to much faster computing technologies, and much harder to break codes which will continue the improvements of computing technologies in our lives.

B) Quantum Teleportation

(14) This term is misleading since it's not teleportation in the "Star Trek" sense. It's all about properties of "entangled" particles being the same over long distances.

The results of Quantum Teleportation are likely more secure messaging and a more secure internet. Security which is much harder if not impossible to break.

Overall, from my research, these technologies should lead to gradual improvements in our overall lives over time, but nothing in terms of a major paradigm shift which will dramatically affect society.

12.16 Robotics

You could argue that the history of Robotics goes all the way back to ancient times...

Concepts akin to a robot can be found as long ago as the 4th century BC, when the Greek mathematician Archytas of Tarentum postulated a mechanical bird he called "The Pigeon", which was propelled by steam.

Yet another early automaton was the clepsydra, made in 250 BC by Ctesibius of Alexandria, a physicist and inventor from Ptolemaic Egypt. Hero of Alexandria (10–70 AD) made numerous innovations in the field of automata, including one that allegedly could speak.

Currently, robots are used in many types of manufacturing includes automobiles and microelectronics. They are much more efficient than humans, don't tire out, and don't need to be paid.

Robots are gradually taking over repetitive tasks in manufacturing and even in the home—like floor cleaning robots. As Robots take on more jobs, you might think this leads to more human unemployment, but the history of the development of technology is that as people are replaced in more mundane work, then new jobs open up which are higher level and need more human insight and directions.

An example are robot programmers. These are humans, and you can't run a robot without programming.

So is there a paradigm shift? Robots seem to be causing a large scale automation change in our lives over time.

You could include automated driving in this long term paradigm shift. I think we will see some robotic changes like robotic cars in the next ten to twenty years.

Another change will likely be minimum wage jobs at fast food places which will be automated using robotics to reduce the number of human employees at these locations.

The long term effects should be an increase in productivity, an increase in human leisure, and more higher level jobs opening up for humans.

It will probably take multiple decades or even centuries for these changes and improvements in our civilization to reach maximum impact.

12.17 Space Travel & Space Settlements

I recently wrote a book titled "Designing and Building Space Colonies - A Blueprint for the Future" so I've done a reasonable amount of research already on this topic.

We will shortly enter an era of commercial and government manned space travel. Both SpaceX and Boeing have been contracted to develop manned space flight technologies to take astronauts to the International Space Station starting in 2019. This will also lead to more commercial ventures in space like Space Hotels which Bigelow Aerospace wants to building using their inflatable habitat technology.

NASA is also forming a partnership to build a "Deep Space Gateway" in orbit around the Moon in the 2020s. There is already a robust satellite building and launching business worldwide today. Over the next few decades we should see addition large investments in space technologies.

This expansion will also include colonies on the Moon and Mars within the next one hundred years.

Interstellar technologies are even being developed including the possibilities of faster than light drives to reach the stars. This includes the potential Alcubierre drive.

The Alcubierre drive or Alcubierre warp drive (or Alcubierre metric, referring to metric tensor) is a speculative idea based on a solution of Einstein's field equations in general relativity as proposed by Mexican theoretical physicist Miguel Alcubierre, by which a spacecraft could achieve apparent faster-than-light travel if a configurable energy-density field lower than that of vacuum (that is, negative mass) could be created. (15)

Space exploration and settlement is a long term paradigm shift and it will take decades, centuries, and thousands of years to fully develop.

I do think all of this exploration and settlement will eventually happen because it is man's nature to explore and settle new locations.

Here is a chapter called "Realities and Reasons" from my book on Space Colonies to give you some ideas of why we should build space colonies:

<u>Realities and Reasons</u>

Space travel is also expensive. This is the biggest limitation imposed on working and living in space. We have the technologies to build more space habitations today, but the cost is prohibitive.

The International Space Station cost about $150 billion dollars to build. This is a huge amount of money. NASA is also budgeting $3-4 billion for continued support through the early 2020s. So we can propose all types of space structures, but the continued cost of space launches will remain the main limiting factor. Until launch costs can be reduced significantly, it will still be difficult to plan things in space.

What SpaceX is doing with re-usable rockets is a real game changer for spaceflight costs.

The physics of building Space Habitats is well understood. It is the development of the technologies and the funding for such structures which is the challenge.

There are also a number of technologies we will require to build large space habitats including replicating robots, replicating three dimensional printers, propulsion technologies, and more. Given all these realities, here are a number of reasons listed in the National Space Society website as to why we should build and live in space habitats:

1)   Proximity to Earth. The first orbital settlements may well be built only a few hundred miles from Earth in 'Low Earth Orbit' (LEO). High LEO is far enough out that the settlement won't crash into Earth but low enough for the Earth's Van Allen Belts to protect settlers from deadly solar storms. Travel back and forth to Earth should take only a few hours. Visits from relatives and friends will be common, and traveling to Earth for vacation or schooling should be easy. Perhaps more important, bringing supplies, materials, and specialized equipment from Earth to support construction will be relatively easy.

2) Continuous, ample, reliable solar energy. In a high enough orbit there is no night. Solar power is available 24/7 in most high orbits, although in high LEO there is some darkness during each orbit as a structure passes through the Earth's shadow. Most satellites in Earth orbit use solar power today, deploying large solar cell arrays like wings stretching from the craft. The solar arrays for settlements must be huge in order to generate enough power. This power can be generated on separate solar power satellites and beamed to the settlement, much as power beamed from such satellites to Earth can play a major role in solving our energy problems.

3) Weightless construction. Zero-g construction means big settlements can be built with relative ease. On Earth, for example, you could not build a round structure that is several miles high because it would collapse under its own weight, but in zero-gravity it is entirely possible to build such large structures, and in orbit astronauts can move objects weighing many tons by hand. Space settlers will spend almost all of their time inside the settlement because it is impossible for an unprotected human to survive outside for more than a few seconds. In this situation, obviously, bigger settlements are better. Settlements can be made so large that, even though you are really inside, it feels like the out-of-doors.

4) Weightless recreation. Although space colonies will have 1g at the hull, in the center you will experience weightlessness. If you've ever jumped off a diving board, you've been weightless. It's the feeling you have after jumping and before you hit the water. The difference in an orbital space settlement is that the feeling will last for as long as you like. If you've ever seen videos of astronauts playing in 0g, you know that weightlessness is fun.

Acrobatics, sports and dance go to a new level when constraints of gravity are removed.

5)   Great views of Earth (and eventually other planets). Space settlement is, at its core, a real estate business. The value of real estate is determined by many things, including "the view." Any space settlement will have a magnificent view of the stars at night. Settlements in Earth orbit will have one of the most stunning views in our solar system: the living, ever-changing Earth.

6)   Enormous growth potential. If the single largest asteroid (Ceres) were to be used to build orbital space settlements, the total living area created would be well over a hundred times the land area of the Earth. This is because Ceres is a solid, three dimensional object but orbital space settlements are basically hollow with only air on the inside. Thus, Ceres alone can provide the building materials for uncrowded homes for hundreds of billions of people, at least.

7)   Economics. Near-Earth orbital settlements can service Earth's tourist, energy, and materials markets. Space settlements, wherever they are built, will be very expensive. Supplying Earth with valuable goods and services will be critical to paying for settlement.

12.18 Transportation

Transportation is still evolving and becoming a lot more intelligent.

a) Flying Cars

Flying cars have been on the drawing boards for decades but never went anywhere. (Some flying cars were even built in the 1940s as airplanes convertible to cars and visa versa.)

Some of the latest flying car designs may really become popular. These cars use multiple vertical computer controlled electric rotors-like many unmanned drones. The advantage of this type of vehicle is that they are all electric the computer does the flying, and they can move in very precise paths.

Here is part of an article on the design:

(16) *The hot new trend in aeronautics these days is VTOL (pronounced vee-tol), which stands for "vertical take-off and landing." These aircraft are known colloquially as "flying cars," but we should probably think of them more as drones crossed with helicopters. Uber is working on a VTOL project, as is Airbus, DARPA, and Google co-founder Larry Page. Because nothing says "I'm very rich and I hate traffic" like a flying car project.*

E-volo, a German aviation startup, has been pursuing ultralight, electrically powered "multicopter" technology for several years now. The company's Volocopter VC200, a 18-rotor drone-helicopter hybrid, took its first crewed flight last year. Today at AERO, Europe's largest general aviation trade fair in Friedrichshafen, Germany, E-volo revealed its first production model: the 2X.

To be sure, the fiber-composite 2X is no aviation powerhouse. It has a maximum range of 17 miles when flying at a speed of 43 mph. It's maximum flight time is 27 minutes at an optimal cruise speed of 31 mph. But if range were no concern, the 2X can fly at a maximum speed of 62 mph.

b) Hypersonic Vehicles

Supersonic and Hypersonic vehicles are being designed and built today. They will allow a new generation of civilian craft to be built and flown which have much lower sonic booms so they can be used commercially by large cities worldwide without the large sonic booms causing damage the old ones used to.

There is a lot of evidence than the United States Military had remote controlled Hypersonic Nuclear Bomb carriers as far back as the 1980s which were designed as a strategic weapon to use against the Soviet Union.

In the next decades we should see more of these vehicles used for commercial and military aviation. This is an aerospace technology trend which is ongoing now.

c) Re-Usable Rockets

A current development of the last few years through now is reusability of rockets from one launch to another. This functionality was pioneered with the SpaceX Falcon 9 rocket where the booster sections fly back to a safe landing after the launch.

Feb 6, 2018 Space X launched a huge rocket called the "Falcon Heavy" with over five million pounds of thrust. For the sample launch it is sending a tesla car to Mars with a mannequin in a spacesuit. The unique thing about the launch was that all three boosters were sent back to land to be re-usable. Two landed properly and the third crashed, but this is still an incredible improvement to the economics of space launches.

Falcon Heavy rockets will be priced at $90 million per launch compared to over $400 million for comparable launches from the United Launch Alliance. (Lower thrust too)

This new rocket reusability and much lower costs will create a huge impact on future rocket designs going forward.

Elon Musk of SpaceX has even bigger plans in the next ten years. His company is designing and plans to launch the BFR or "Big Falcon Rocket" which will be more powerful than the record holding Saturn V rocket and which will reduce the cost of launching materials into orbit to lower than any other launch platforms.

I see this trend continuing for the next 200 years. Other launch companies will either adapt or die. Long term the costs of space access will continue to fall.

d)   Space Elevators

Space Elevators were postulated in 1895 by Konstantin Tsiolkovsky. His proposal was for a free-standing tower reaching from the surface of Earth to the height of geostationary orbit. Like all buildings, Tsiolkovsky's structure would be under compression, supporting its weight from below.

Since 1959, most ideas for space elevators have focused on purely tensile structures, with the weight of the system held up from above by centrifugal forces. In the tensile concepts, a space tether reaches from a large mass (the counterweight) beyond geostationary orbit to the ground. This structure is held in tension between Earth and the counterweight like an upside-down plumb bob.

We still need to develop materials strong enough to build a space elevator and find how to provide the funding to build one, since it would probably cost at least hundreds of billions of dollars for construction.

The big benefit would be the incredibly low costs for space access. On the order of a few dollars per pound versus the thousands of dollars per pound today.

This could be a huge paradigm changer but is not likely to happen for a couple of hundred years.

13.0 Human Intelligence

One subject is so critical to our future that it needs to be looked at separately: I'm talking about potential improvements in human intelligence from genetic engineering and integration of computers.

The common belief is that genetic engineering can create embryos which will all be super geniuses. How will the world change if everyone is a genius? And what is a genius? Are we talking somebody who can solve incredible math problems in their head or somebody like Leonardo Da Vinci who had incredible creativity?

I think this belief illustrates a fundamental misunderstanding about human intelligence. Here is a quote which partially illustrates what I'm talking about:

> *A smart man makes a mistake, learns from it, and never makes that mistake again. But a wise man finds a smart man and learns from him how to avoid the mistake altogether.--- Roy H. Williams*

My belief is that our being is both physical and non-physical. I.E. that our consciousness is both in the brain and in the spirit. This being the case, then there are limits to how much effect improving the brain can improve consciousness.

If everyone has a photographic memory and top reasoning ability, this still doesn't necessarily have an impact on creativity and artistic abilities.

(17) The standard I.Q. tests measure mostly memory and reasoning abilities. There are a few tests which also try to measure creativity, but how do you measure something which nobody really understands?

So what is the impact of massively increased intelligence on society? Genetic engineering could make it such that every baby born has a perfect memory and incredible reasoning capabilities.

However, while this can help creativity, it doesn't mean that creativity and other qualities of consciousness will automatically be produced by this higher intelligence.

My thoughts about the impact of genetically improved intelligence on society is that there are real limits. You can improve the brain to become a much more efficient machine to recite facts and improve reasoning.

However, this is only part of overall intelligence. What about these factors which make up our total being consciousness:

- Emotional Intelligence
- Leadership
- Creativity
- Energy to do Things
- Persistence
- Purpose In Life
- Wisdom
- Spirituality
- Compassion

You can modify human genetics massively towards improved intelligence, but I don't know that giving people a perfect brain will significantly affect the qualities above.

There is also some work being done to create biological computers which might be implanted in human beings. This would dramatically increase memory and possibly logical reasoning. However, there would still be the same limits on overall intelligence as discussed earlier.

My overall conclusion about intelligence as a factor changing our future is that people will become smarter over time as their brains are improved through genetics, but it will not create a massive paradigm shift in our world as some people look forward or even fear.

"Limitless" is one of my favorite movies. It's about a man of above average intelligence who takes a pill which maximizes his brain usage and he becomes the smartest man in the world. He has a massive impact on the world as a result.

Although I love the movie, I don't feel it is realistic, since just boosting the brain doesn't provide all of the qualities of consciousness as listed above.

As a result I do not include improvements in intelligence in my predictions. We will have some improvements from genetics and linkages to computers but I don't see any radical paradigm changes.

Future Predictions By an Engineer and Seer

14.0 Trends in Human Interactions

Besides technology there are many other important trends in government, human society, companies, work, and more which we need to examine.

14.1 Overall Trends in Work and Jobs

Today many millions of people work part time, remotely, on contract, and generally are independent contractors.

Freelance Job sites like "Upwork.com" connect clients with skill providers worldwide. What is really amazing from twenty years ago is that with the internet you can hire people anywhere in the world to do all sorts of tasks for your business and get the best rates available.

I've used many of these contractors for my business from Africa to India, and they have reduced the costs to my business significantly.

This trend should only continue as more people get online worldwide, and as even more skills can be performed remotely due to the increasing power of the internet.

Company's regular employees will work more and more with contractors and freelancers as the work environment matures into using the correct and changing blend of these resources to accomplish all types of jobs.

More usage of virtual reality will continue to support the growth of freelancers and contractors in remote areas since this will make it much easier for them to interface with full time employees and customers.

Additionally, companies will offer more health and lifestyle benefits and options to keep good employees and find and hire new ones.

I see these trends continuing into the foreseeable future with remote work becoming more and more prominent but not a dramatic and immediate shift in work life, styles of work, or locations. The current trends will just keep going.

14.2 Trends in Corporations

The growth in companies large and small will continue.

Since the invention of the internet and mobile phones there are now small businesses in underdeveloped countries around the world using them.

Examples would be fishermen in poor countries in Africa who use mobile phones to find out which nearby cities will give them the best prices for their fish.

Or poor women in India who rent cell phones to people in their villages. This can be very profitable to them. There are even banks in those countries which make micro loans to help them purchase the cell phones in the first place.

Almost every country in the world now realizes the importance of free enterprise and business to help their economies grow. The only countries counter to this trend are communistic regimes like in North Korea—and those places are hell holes. Another heavily socialistic "paradise" is Venezuela which is falling apart more every day and is close to revolution.

The development of businesses through new ideas and technologies has proven so successful that there are even large venture funds and online capital raising to help these creative people to create new businesses.

Overall, businesses and capitalism seems very healthy now and into the foreseeable future. I don't see that changing except in good ways.

It is possible that in the future almost everyone will be their own consultant or freelancer offering their skills to others over the internet.

14.3 Trends in Government

Since 1945 the number of countries in the world has increased significantly. If we go by the memberships in the United Nations we had 51 members in 1945 and over 193 in 2011.

This shows that countries have fragmented radically into almost four time the number of them worldwide over the last seventy three years. It's possible that the number of countries will continue to grow as more countries fragment from social, economic, and political conflicts.

So what does this fragmenting mean to the average Joe on the street? Most probably don't worry about it since it doesn't change the government services offered very much—whether these services are good or bad. And many products or services they buy are from major corporations around the world anyway.

Overall, this fragmentation may give people in these new countries more of a political identity which they like.

14.4 Trends in Religious and Spiritual Areas

In today's worldwide society several long range trends exist:

Traditional Religious denominations in Christianity are losing interest among the young. In the United States Presbyterians, Lutherans, Methodists, and similar denominations are shrinking while mega-churches and fundamentalist sects are growing—many with almost all young people in their twenties as participants.

A Worldwide Demographic study from 2015 says (18):

The religious profile of the world is rapidly changing, driven primarily by differences in fertility rates and the size of youth populations among the world's major religions, as well as by people switching faiths. Over the next four decades, Christians will remain the largest religious group, but Islam will grow faster than any other major religion. If current trends continue, by 2050…

- The number of Muslims will nearly equal the number of Christians around the world.
- Atheists, agnostics and other people who do not affiliate with any religion – though increasing in countries such as the United States and France – will make up a declining share of the world's total population.
- The global Buddhist population will be about the same size it was in 2010, while the Hindu and Jewish populations will be larger than they are today.
- In Europe, Muslims will make up 10% of the overall population.
- India will retain a Hindu majority but also will have the largest Muslim population of any country in the world, surpassing Indonesia.
- In the United States, Christians will decline from more than three-quarters of the population in 2010 to two-thirds in 2050, and Judaism will no longer be the largest non-Christian religion. Muslims will be more numerous in the U.S. than people who identify as Jewish on the basis of religion.

- Four out of every 10 Christians in the world will live in sub-Saharan Africa.

So we can see that religion will continue to play as big a part in world society later in the twenty first century as it does now.

Islam is still growing almost twice as fast as Christianity and will have the same number of adherents in the mid twenty first century.

Some areas like Europe will have a much higher percentage of their population who are Islamic which will cause political directions to sway more in that direction.

Otherwise I don't see a huge shift in the religious beliefs of people projecting into the future.

One issue to watch carefully is the growth of Islamic Terrorism. This has been a big problem for the world in the last few decades. However, some countries which used to be the biggest funders of extremist ideologies like Islamic terrorism seem to be backing off their support as they have realized that these extremist groups are now attacking their own societies.

While I think Islamic Terrorism will still be with us in the future, it seems to have reached its high water mark worldwide and is receding.

My perspective is that of a person who is a pure "Spiritualist" who doesn't follow any direct religion but believes in our ability to directly connect our consciousness with God through meditation and similar practices. It looks like people in the world who think like I do will continue to be a very small minority.

14.5 Trends in Social Interaction and Relationships

One trend which is already going on in the United States and in other countries is more acceptance for alternative lifestyles such as Gay, Lesbian, or related lifestyles. Gay marriages are now accepted in all of the United States.

In some very conservative countries like Saudi Arabia women are getting more and more rights like the right to drive.

What other types of relationships might develop?

Younger people are spending a huge amount of their time in communicating with each other on social media. This has already happened and I think they have invested a maximum of their time in these "Electronic Relationships".

So "Electronic Relationships" will continue to be a large part of everyday life and may even become more intimate as technology advances.

Science Fiction has projected than humanity might evolve many new types of marriages in the future such as contract marriages for a defined term, or multiple marriages with several to many men and women all in one marriage. This does have advantages for group care of and raising children.

However, I don't see any indication of these new types of marriages developing at this time. Even the Mormons who were historically all for one man and multiple women marriages don't seem to be pushing those types of social relationships anymore.

Overall, I don't see any additional huge evolutions in social relationships going forward. Maybe there is a hidden Paradigm shift to look out for?

15.0 Paradigm Shifting

When predicting the future there are some variables which are hard to predict:

a) Will certain trends continue and how long will they take? We might all agree in the 1970s that CPU densities were going to increase for a number or years but who would have guessed that this trend would continue until now—almost half a century later!

b) When will a major Paradigm shift occur and can we even see it coming? We can guess at what major shifts will be headed towards us but might we not see them at all? The impact of the Internet integrated with Smartphone usage wasn't thought of by hardly anyone.

15.1 Summary of Paradigm Shift Potentials

Before I get into actual future predictions I want to nail down potential paradigm shifts as much as possible.

As I've reviewed in earlier chapters, the future is a mix of projections of different trends and paradigm shifts. Many paradigm shifts are by their natures never thought of in advance. These shifts cause the largest and most dramatic changes in future events. This is why I'm putting so much effort into defining them.

It's also possible that many trends will intersect to create unexpected major changes in our lives. A good example as I've related in previous chapters has to do with how the Internet has intersected with GPS Navigation, Smartphones, and Digital Media content to make major changes in our daily lives.

Technology Driven Probable Paradigm Shifts:

Segment	Paradigm Shift Type	Description	Timeframe
Biology	CRISPR Technology	Large Scale Disease and Biology changes	Now-200 Years
Digital Currency	Crypto Currencies	Digital Currencies independent of any state	Now to 100 Years
Education	Remote Learning	A gradual shift towards most learning online	Now to 100 Years
Farming	Hydroponics	Large scale increased food production	Now-100 Years
Power	Fusion Power	Fusion Power and then miniaturization	30 years-300 years
Health & Medicine	Increasing Longevity	Personal Processes and Medicine to continually improve longevity	Now-1000 years
Intelligence	Human Intelligence	Genetic Engineering to dramatically improve human intelligence	25 Years to 200 Years
Manufacturing	3D Printing	3D Printing in all phases of manufacturing	Now-300 Years
Military	Nuclear Proliferation	Small Unstable Powers and small Nuke Wars—hypersonic aerospace vehicles	Now-150 Years
Nano Technology	Nano Robots	Nanotech used for medicine, war, construction, and more	Now-500 Years
Robotics	Robotics Of All Types	Replacing repetitive tasks and construction, including uses at home	Now-250 Years
Construction	Nest Cities	Horizontal Elevators and all types of movements in cities	Now-300 Years
Space	Space Travel & Colonies	Exploration and Settlement	Now-10,000+ Years
Transportation	Flying Cars	Cars using electric rotors and computer control for personal transportation and Hypersonic airplanes.	5 Years to 100 Years
Transportation	Reusable Rockets	Happening now and continued reductions in cost should continue	Now-200 years
Transportation	Space Elevator	Building a Usable Space Elevator	200-1000 years

Far Out Possible Paradigm Shifts:

Focus	Paradigm Shift Type	Description	Prob	Time Frame
Gravity	Anti-Gravity	Use of Anti-gravity technology for transportation and in phases of society	80%	50 Years to 200 Years
Paranormal	Mass Psychic Abilities	That Psychic abilities will be generally accepted and people trained in their usage	95%	Now to 10,000 Years
Zero Point Energy	Zero Point	Mass Zero Point Energy production from the vacuum	10%	500 Years

Future Predictions By an Engineer and Seer

Future Predictions By an Engineer and Seer

16.0 The Speed of Change

An important observation has come to me as I reviewed expectations of changes and improvements in technology over the centuries.

Many of these changes will happen gradually instead of all at once. It may take hundreds of years for some of these new technologies to be fully mature.

Good examples are improvements in power generation which might take hundreds of years for their full effect to impact all of us.

Another area is 3D manufacturing. We probably haven't even thought about some of the things which can be done with 3D printing given the improvements in materials and designs to come.

Also, there might be a mergence of certain technologies in the future.

What about the mergence of nano tech and 3D printing? Building things automatically from the Atomic level.

Or people having integrated links to the Internet built into their brains from biological improvements and improved wireless networking. Think about having the worldwide access to all information on call just by thinking about it.

What about a spaceship or space colony built entirely by Robots and 3D printing? All you have to do is supply the materials. This is a theme in my book "Personal Freedom 1&2 Building a Space Colony and A Trip to the Stars".

My point is that many of the coming paradigm changes may already be in process but are so gradual that we might not see the effects within our lifetimes. This doesn't reduce the importance of these changes to humanity even though they may occur in very long timeframes.

17.0 Predictions About Human Civilization

Now it's time to make actual predictions on where humanity will be at different dates in the future.

Again, the following predictions of life in the future are a combination of my research into trends, defined potential paradigm changes, and some intuition.

What I've tried to do for each set of predictions is to describe how an individual or family will live in the designated time with the future changes that we expect and/or how civilization will be in general at that point in time.

17.1 In the Year 2100 AD—Roughly 80 Years

Here is a typical day in the year 2100 AD....

My alarm woke me and I got ready for work this morning. I showered and cleaned up, then turned on my home 3D printer to print a new dress shirt. I was tired of the old one. The new one is fresh material and has some nice prints on it.

Most days I work at home, but today I wanted to be at the office because I am going to meet my wife at her Doctor later this afternoon. I also want to just say high to some friends at work since I don't get in there very often.

I went into the kitchen of my apartment and got a frozen breakfast out of the freezer to heat up in the microwave. Kissed my wife goodbye since she was up and headed to the garage.

Time to leave for work so I got into my Ford Electro flying car for the one hour commute into the city. As I belt in I confirm the program to take me to my downtown office from my suburban location. The car rolled out of the garage and then the twelve electric rotors spun up and we went up into the air.

The flying car navigated using GPS to the nearest public airway and we flew with traffic separated every one hundred feet into the city. When we got to the parking lot near my building we landed and the car automatically taxied into the underground lot.

I got out and went to the nearest intelligent elevator. I put in my destination as did a few other people in the car. The doors closed and the software navigated to the closest offices. We went up several floors, then the elevator moved sideways into the building across the street. Then it went up to various floors to drop everyone off.

Walking to my office I got online for this morning's virtual meeting. I wore my virtual viewing goggles and found myself in the virtual conference room with my peers for our morning marketing meeting. We all talked to each other like we were in the same room and presentations were displaying in the air on a screen which seemed to hover in the air.

After the meeting I had time to call my wife. She and I confirmed I would meet her at the doctor's right after lunch.

My flying car took me to the medical complex which was about five miles away and I had a few minutes to just walk around before Janet got there.

This complex of buildings had been built in the last ten years and had many connections between buildings—above and below ground for the intelligent elevators.

Janet and I met on time at the Doctors office. She had been diagnosed a couple of weeks ago with Lou Gehrig's disease. It used to be a fatal long term disease where suffers experienced lessened muscle strength over years until they couldn't even breathe anymore.

I say used to but with modern genetic manipulations from a descendant of the CRISPR technique these genetics diseases could be modified to totally remove them people's systems.

Janet just need to have an infusion into her system from a bag with a needle into her veins which took an hour. Then she was cured.

That evening we contacted her brother Sam to give him the good news since he had been worried.

He was an Astronaut Astronomer and lived in the Moon colony at Shakelton Crater in the South Pole of the Moon. There were only twenty persons there at any one time. The colony was an extension of the 2020s Deep Space Gateway in orbit around the moon. There was a larger station there now serving both moon access and where the first few expeditions had left from to visit Mars.

Future Predictions By an Engineer and Seer

Sam talked to us on a video link and it was delayed a few seconds since the signal had to go all the way to and from the Moon.

We asked Sam how he spent his free time and he said he was working on a new Software Engineering Degree in the online Stanford University School. They had online classes and he could also do remote consultations with his instructors too. He said he would be finished with the B.S. degree in another year.

Janet and I also did our daily wellness practices that evening which consisted of some Yoga, and Meditation, to help us live decades longer. The average lifespan in the United States was now over 100 mainly due to most people adopting these wellness practices.

We ended the call and went to the airport the next morning to pick up our daughter Gina who was visiting from Australia where she was working. She was taking the Hypersonic plane from there and the sub-orbital trip would take her less than one hour to go from Sydney, Australia to the International Airport in Boston where we were picking her up.

While my wife was making dinner I watched the world news on the wall screen TV and saw a report on a war in Africa where lasers and orbital bombardment weapons were being used to attack the rebels.

Future Predictions By an Engineer and Seer

17.2 In the Year 2250 AD—Roughly 232 Years

What life is like in the mid twenty third century….

Technology keeps changing in our lives century after century to make life more comfortable. The Earth stabilized at about 12 billion people over one hundred years ago in the mid twenty second century. Most people studying population trends believe that the higher standard of living and the incredible increases in hydroponically grown food led to worldwide population stability.

It was no longer economically worthwhile to have more than two children, because of the cost of raising them and educating them, so people self-limited themselves generally.

We were fortunate with the incredible developments of hydroponics technology which had such great food production that it was scaled up and able to feed the entire Earth's population. Today all of this hydroponics technology is a big part of successful settlements in Space.

The biggest technology change in life in the last couple hundred years has been that anti-gravity is now used all over our civilization along with cheap fusion power.

The use of relatively inexpensive anti-gravity technology has changed lives in many ways. To start with, the cost of bringing materials into Earth Orbit have gone down by several factors of ten. This means it now costs no more to take a truck load of things into orbit as it does to drive to the nearest city one hundred miles away.

With the reduced costs of materials transported into orbit from Earth and new technologies in Space, exploration of the solar system and space industries have flourished.

We now have space hotels where people go on vacation instead of the Caribbean or Europe. Space tourism is a one hundred billion dollar business.

We are mining asteroids and smelting materials to get metals at much lower costs than they could be mined on Earth. The Martian Colonies are close to one hundred and the largest city has over one million people in it. They still use domes and underground habitations there since any terraforming of Mars will still take centuries to complete.

Small Habitats are being experimented with in near Earth Space but the amount of money to build them is still an issue.

On Earth, anti-gravity transportation has replaced most other forms of transport because moving things by anti-gravity with cheap fusion powered thrusters is just a lot cheaper than other modes of transportation.

In terms of an improved humanity, many parents program the qualities their babies will have ahead of time. This includes high athletic abilities and high intelligence. In fact there are now popular "Baby Templates" for parents to choose from which emphasize different career expertise or personality types.

Buy and selling is now done using one of multiple Crypto currencies. Money is no longer controlled by governments since government currencies lost all credibility with their currencies back in the 2110 crash of world markets.

Now you may have credit with several Crypto currencies depending on what you are buying or selling. There are different ones for business and personal usage, as well as transportation or retail purchases.

Living offshore on the oceans has become very popular and there are over one hundred cities on the oceans. The largest is over 100,000 people. These cities support themselves mainly through the aqua culture of fish and mollusks.

17.3 In the Year 3000 AD—Roughly 980 Years

The year 3,000 A.D. is a milestone for Humanity. Many people in the past thought we would exterminate ourselves from the Earth many centuries ago. Instead Mankind is flourishing.

With the Solar System being fairly extensively settled, there is now no danger of Man going extinct except from a huge interstellar War or some type of Supernova in a nearby star.

We also finally started building large O'Neil settlements in Space. They each consist of two counter rotating cylinders five miles in diameter and twenty miles long, each capable of holding over one million people. (The designers decided to stick to rotation for gravity since anti-gravity devices for this large a colony would suck too much power.)

The economic barrier to make these colonies profitable was crossed when asteroid generated materials costs fell to pennies, and advanced robotic and 3D assembler technologies allowed these huge archologies to be built in almost an entirely automated manner in space.

We now have several colonies in near Earth Space and several others around the Solar System including at Mars, the Asteroid Belt, and further out to Neptune.

In the last hundred years we launched a series of unmanned probes to the nearest stars using the experimental faster than light Alcubierre drive. Development of this drive started in the twenty-first century and it actually creates a space bubble which puts the spaceship into a frame of reference where it can exceed normal light speed. It still takes a month to go a light year.

Our probes had high end Artificial Intelligence and reported back that over ten planets so far had life on them which we could settle.

Our first Interstellar Colony spaceship is being built and it's a modification of an existing space colony or O'Neil Habitat.

The plan is that it will leave the Solar System in ten more years to plant man's first colony on an extraterrestrial planet at another star!

On our home planet Earth, the cities now look like incredibly complex designs. A combination of several mile high towers and intelligent elevators which move between buildings both horizontally and vertically have led to a city center which looks much more like a complex system of connections that an anthill would have—but above ground.

Every home now has a very sophisticated building machine which can build almost anything we need from scratch. All you need to do is put in raw materials. These machines use nanobots to build everything from a microscopic level up to full size objects. So you can build all sorts of electronic devices from the atomic level up. Some of the things we build for ourselves at home include food for meals, clothes with embedded electronics for our smart devices and all sorts of things for the home.

We don't use washing and drying machines anymore since we can create new intelligent clothes for ourselves every morning.

The average lifespan is now 300 years and continues to increase. This means that everyone can have multiple careers and switch families as their lives change. Contract marriages with fifty year limits are now common.

Now that everyone has perfect memories and incredible logic from our DNA programming before birth, education is a much faster process.

But, childhood has now been extended to almost fifty years as young people learn what they want to do with their lives after experimenting with multiple lifestyles and careers in their younger years.

18.0 Summary

I think we will continue to have some disasters on Earth but overall, humanity should continue to advance and occupy space over the coming centuries.

Oh, there will be small nuclear wars, major Earthquakes, even large Volcanos going off, but humanity will overcome all of these problems.

The Universe is about 13.8 Billion years old and many theorists predict the Universe may last 100 Trillion years. If this is true, then humanity is at the barest beginning of life in the Universe. Humanity has a great future ahead and we may end up being the wisest and oldest race in the Universe. Think about that.

Marty Ettington

February 2018

Bibliography

1. Smatterist.com. *http://smatterist.com/6289/18-predictions-from-100-years-ago-that-didnt-come-true-but-still-sound-amazing/.* [Online] 2018.

2. Climate, Sunspots and. http://www-das.uwyo.edu/~geerts/cwx/notes/chap02/sunspots.html. *http://www-das.uwyo.edu.* [Online] 2018.

3. Strohmeyer, Robert. https://www.pcworld.com/article/155984/worst_tech_predictions.html. *https://www.pcworld.com.* [Online] 2008.

4. Theory, Malthusian. https://en.wikipedia.org/wiki/Malthusianism. *en.wikipedia.com.* [Online]

5. Walter, Damien G. https://damiengwalter.com/2014/11/14/can-you-name-arthur-c-clarkes-top-5-astounding-predictions/. *damiengwalter.com.* [Online] 2018.

6. True, Futuristic Predictions of H.G. Wells That Came. https://www.smithsonianmag.com/arts-culture/many-futuristic-predictions-hg-wells-came-true-180960546/. *https://www.smithsonianmag.com.* [Online] 2016.

7. Law, Moores. en.wikipedia.org. *https://en.wikipedia.org/wiki/Moore%27s_law.* [Online] 2018.

8. Growth, World Population. https://en.wikipedia.org/wiki/Population_growth. *en.wikipedia.org.* [Online] 2017.

9. CRISPR. https://www.livescience.com/58790-crispr-explained.html. *livescience.com.* [Online] 4 17, 2017.

10. Cryptocurrency. https://en.wikipedia.org/wiki/Cryptocurrency. *https://en.wikipedia.org.* [Online] 2018.

11. Hydroponics. https://home.howstuffworks.com/lawn-garden/professional-landscaping/alternative-methods/hydroponics7.htm. *howstuffworks.com.* [Online] 2018.

12. Density, Battery Energy. https://www.elektormagazine.com/news/new-cell-technology-3-or-15-times-today-s-energy-density. *www.elektormagazine.com.* [Online] 2018.

13. Medicine, 10 Exciting Improvements in. http://medicalfuturist.com/10-exciting-medical-technologies-that-will-make-you-hopeful-about-our-future/. *http://medicalfuturist.com.* [Online] 2018.

14. Teleportation, Quantum. https://www.nasa.gov/feature/jpl/teleporting-toward-a-quantum-internet. *https://www.nasa.gov.* [Online] 2018.

15. Drive, Alcubierre. https://en.wikipedia.org/wiki/Alcubierre_drive. *https://en.wikipedia.org.* [Online] 2018.

16. Cars, Flying. https://www.theverge.com/2017/4/5/15195786/evolo-volocopter-2x-vtol-flying-taxi-announcement. *https://www.theverge.co.* [Online] 2018.

17. creativity, intelligence and. https://creativesomething.net/post/41103661291/the-relationship-between-creativity-and. *https://creativesomething.net.* [Online] 2013.

18. Religions, The Future of World. http://www.pewforum.org/2015/04/02/religious-projections-2010-2050/. *http://www.pewforum.org.* [Online] 2015.

www.ingramcontent.com/pod-product-compliance
Lightning Source LLC
Chambersburg PA
CBHW031415210526
45464CB00005B/1898